CAMBRIDGE COUNTY GEOGRAPHIES
SCOTLAND
General Editor: W. Murison, M.A.

EAST LOTHIAN

T0346147

Cambridge County Geographies

EAST LOTHIAN

by

T. S. MUIR, M.A., F.R.S.G.S.

Geography Master, Royal High School, Edinburgh
Author of *Edinburgh and District*, and *Linlithgowshire*

With Maps, Diagrams and Illustrations

Cambridge:
at the University Press
1915

CAMBRIDGE UNIVERSITY PRESS
Cambridge, New York, Melbourne, Madrid, Cape Town,
Singapore, São Paulo, Delhi, Mexico City

Cambridge University Press
The Edinburgh Building, Cambridge CB2 8RU, UK

Published in the United States of America by Cambridge University Press, New York

www.cambridge.org
Information on this title: www.cambridge.org/9781107637931

First published 1915
First paperback edition 2013

A catalogue record for this publication is available from the British Library

ISBN 978-1-107-63793-1 Paperback

CONTENTS

CONTENTS

ILLUSTRATIONS

ILLUSTRATIONS

MAPS

NOTE.—Thanks are due to H.M. Geological Survey of Scotland for the illustrations on pp. 26 and 29; to the Society of Antiquaries of Scotland for those on pp. 79 and 80; to James Curle, Esq., W.S., for that on p. 82; to the Earl of Haddington for indicating the source of that on p. 84 and to Mr J. Spence, Musselburgh, for the photograph; to Messrs G. P. Putnam's for permission to reproduce the portrait on p. 105 from Cowan's *John Knox*; and to Mr Charles E. Green for permission to reproduce those on pp. 77 and 90 from his *East Lothian*.

The illustrations on pp. 4, 7, 12, 14, 16, 18, 35, 38, 39, 41, 45, 60, 63, 66, 69, 86, 88, 93, 99, 111, and 112 are from photographs by Messrs J. Valentine and Sons, Dundee.

1. County and Shire. The Origin of Haddingtonshire or East Lothian.

The word *shire* is of Old English origin and meant office, charge, administration. The Norman Conquest introduced the word *county*—through French from the Latin *comitatus*, which in mediaeval documents designates the shire. *County* is the district ruled by a count, the king's *comes*, the equivalent of the older English term *earl*. This system of local administration entered Scotland as part of the Anglo-Norman influence that strongly affected our country after the year 1100. The earliest mention of a sheriff of Haddington is in the thirteenth century. Although the county boundary, except along the sea-board, is almost nowhere a natural one, there is every reason to believe that it has remained nearly unchanged for centuries.

About half of the counties of Scotland are named after the county town. Officially, i.e. in census and other government returns, our county appears as Haddingtonshire ; but its colloquial designation is East Lothian. The origin of the name Haddington is unknown. The earliest spellings are Hadynton (1098), and Hadintune, Hadingtoun, in the twelfth century. It is obviously

English, and most probably derived from the founder. The origin of the word Lothian is less obscure. Bede has *Regio Loidis*, i.e. the same as Leeds. In the Pictish Chronicle, year 970, it appears as *Loonia* ; a century later in the Old English Chronicle, it is *Lothene*. In the Welsh Pedigrees of the Saints is mentioned a prince Leudun Luydawc from the fortress of Eidyn in the North, and the district round Eidyn is named Lleuddiniawn, which, according to the late Sir E. Anwyl, is like a Welsh derivative of Laudīnus. This prince Leudun or Lleuddin is the same whom Sir Thomas Malory calls King Lot. As a territory Lothian originally included all the country to the east of Strathclyde between the rivers Forth and Tweed.

2. General Characteristics. Position and Natural Conditions.

East Lothian unites in itself two of the three physical divisions of Scotland—the Central Plain, and the Southern Uplands. It is also a maritime county, presenting one side to the Firth of Forth, another to the North Sea. Though the scenery has nowhere pretensions to grandeur, yet the coast at Dunbar is bold and picturesque, the Lammermuirs are characterised by massiveness and sweeping curves, while between is the somewhat monotonous but rich and fertile plain, the garden of Scotland. When the traveller from England by the East Coast Route, after descending to Cockburnspath, enters East Lothian, and is whirled past Dunbar, East Linton, and Prestonpans, even the most cursory glance shows him that here is the home of high farming. The huge fields, trim hedges and fences, the rarity of woods, of uncared-for pieces of ground, and of open drains, are ample evidence that the soil is too valuable to permit of any waste.

Down by the sea are dotted little towns and villages, in some cases continuous, in others separated by wide stretches of sandy links. Most of the male inhabitants are engaged in fishing, though in the west round Preston-pans the coal mines claim their servants. The golf-courses of the north-east are second in fame only to that of St Andrews. The burghers of trim North Berwick are as used to visitors of royal and exalted blood as are the citizens of the most favoured of capitals. During the summer months North Berwick and its neighbours are mere suburbs of Edinburgh.

Through the heart of the county meanders the Tyne, famed as a trouting stream. It is emphatically the river of East Lothian, and on its banks and on those of its upland tributaries are many spots celebrated in history and tradition. Ordinarily a placid stream, the sudden melting of the snows or a " cloud-burst " in the Lammer-muirs now and again causes a flood by which the old folks keep their calendar.

To the south of the Tyne the land slopes gently upwards. On the way, however, one comes to sudden and unexpected descents and ascents. These are valleys which run almost parallel instead of perpendicular to the

contour lines as all proper valleys should. For their
origin we must go back to the Great Ice Age. In them
nestle villages and hamlets renowned for their sylvan
beauty, each with its pretty church embosomed in trees
and surrounded by fertile fields. Here and there are
pleasant mansions, the successors of the peels which

The Bass Rock

guarded the passes to the south. But a little way from
their wooded parks begin the bare Lammermuirs. South
of the watershed the East Lothian portion is as bleak,
lonely, and inaccessible as the northern slopes and plains
are fertile, populous, and frequented.

Rising from the plain are several eminences similar in
origin to those other rocks farther west between Edinburgh

and Dumbarton. The Garleton Hills, Traprain, and North Berwick Law are masses of igneous rocks which have withstood the levelling influences of denudation and thus pleasingly diversify the landscape. The Bass Rock, three miles out from Canty Bay, is the boldest of them all.

Despite its mining and fishing, East Lothian is mainly an agricultural county. The rural population out-numbers the urban. No industries exist of other than local importance. It would seem, therefore, that its story should be short and featureless. Yet in agriculture it has been and is an example to the world; its rocks have provided object-lessons and arguments to divergent schools of geology; it has churches and castles of intense interest to the architect and the antiquarian; some of Scotland's most famous sons were born within its bounds; and in history it yields to few districts for importance and romance.

3. Size. Shape. Boundaries.

East Lothian is twenty-fifth in order of size among the counties of Scotland. Only eight are smaller. The exact area is 170,971 acres, or, including water, 171,389 acres—rather less than 270 square miles. It forms a little more than the one-hundred-and-eleventh part of Scotland. Midlothian is more than one-third larger, West Lothian is less than half the size. The shape is an irregular quadrilateral with two long and two short sides. The long sides face south and north-east, the short ones face north and west   On the north-east is the North Sea, on the north-west the Firth of Forth. The western boundary strikes irregularly across country from the Firth to the Lammermuirs. The southern boundary encloses practically the whole of the higher regions of these hills, and finally follows a burn to the sea. The coast, extending between Levenhall and the mouth of the Dunglass Burn, is 32 miles. The longest straight line that can be drawn in the county is one of 26 miles from east to west. The breadth between north and south along a meridian varies from 12 miles in the west to 16 in the middle and 10 in the east. A walk along the boundary, including the coast and all irregularities, would mean a distance of about 80 miles.

On the west is Midlothian, on the south Berwick-shire. Beginning at the mouth of the Dunglass Burn the boundary follows that stream up to the confluence of the Oldhamstocks Burn. Then it keeps to a tributary, the Berwick Burn, and, reaching a narrow plantation called the Dod Strip, follows it to the summit of Dod Hill (1147 feet) Descending to the south-west, it crosses the Eye, and turns sharply south-east, east, and south to the Monynut Water. It then runs up the Philip and comes down to the Whiteadder. The general direction is then westward. At one point it touches the watershed between the Dye and the Faseny, both tributaries of the Whiteadder, itself a tribu-tary of the Tweed. Here the boundary passes over the summit of Meikle Says Law (1750 feet), the highest of

Where the Lammermuirs reach the sea

the Lammermuirs. Near Lammer Law (1733 feet), the highest of the Lammermuirs which is wholly in East Lothian, it follows for a time the water parting between the Tyne and the Leader, a tributary of the Tweed. At West Hill (1479 feet) it leaves the watershed, this time for the north side, and touching the Dean, the East Water, and the Salters Burn, strikes N.N.E. to the Tyne. Thence, keeping the same general direction but with apparently capricious deviations, it reaches the Firth at Levenhall, between Musselburgh and Prestonpans. The numerous islets off the coast, including the Bass Rock, form part of the county.

From this description it will be seen that the land boundary is nowhere determined by geographical features. The simplest explanation is that there are no striking geographical features of sufficient importance to form inevitable boundaries. The Lammermuirs are no Pyrenees, but a much-dissected plateau, and do not present anywhere a well-defined ridge. It is true also the higher parts are entirely given over to sheep ; and, as the inhabited portions of East Lothian are nearer the hills than are the corresponding areas in Berwickshire, it seems natural that Haddington should overlap the watershed. On the western side the boundary follows uncertainly the Roman Camp Ridge, an elongated dome of low elevation, which separates the coal-basin of Mid from that of East Lothian. The probability is that the boundary here was determined by the limits of the farms.

The Commissioners under the Local Government Act

of 1889 made several rectifications of the boundary line. The parish of Fala and Soutra, previously partly in Midlothian and partly in Haddington, was all given to Midlothian. A detached portion of the parish of Humbie was now handed over to Soutra and to Midlothian. A portion of Oldhamstocks parish which had been in Berwickshire was transferred to East Lothian, in exchange for a detached part of Oldhamstocks in Coldingham. The origin of such detached portions is almost invariably ecclesiastical. Some feudal lord endows a church or an abbey with lands from his estates, which may be widely separated.

4. Surface and General Features.

Since the Lammermuirs stretch from south-west to north-east, the part of East Lothian north of the watershed might be expected to drain to the Firth of Forth. The general slope is certainly from south-east to north-west, but the portion which actually belongs to the basin of the Firth is insignificant in area.   This results from the configuration of the county, which we shall now proceed to examine.

A line drawn from the south-west to the north-east corner of the O.S. one-inch Map, Sheet 33, coincides very closely with the great Lammermuir fault, which separates the Lowlands from the Southern Uplands of Scotland.   This line divides the county into two very nearly equal parts, which, with an exception to be mentioned presently, are perfectly distinct.   To the

south of Dunbar, and, therefore, to the south of the line,
a cross-fault, running from north-west to south-east, has
allowed the Lowlands to encroach upon the Lammermuirs, and partially to cut them off from the sea. Oddly
enough, this usurpation is covered with rich soil, forming
part of the famous "red lands" of Dunbar. Geographi-
cally, then, East Lothian is partly in the Southern
Uplands, and partly in the Central Valley of Scotland.

The Lammermuir Hills are the north-eastern portion
of that plateau of ancient rocks which extends across
Scotland from Wigtownshire to St Abbs Head. That it
is a plateau, although much denuded and dissected, will be
realised from a study of the map. Round Meikle Says
Law is an area of more than five square miles which has
no contour line below 1500 feet. In it are marked
eleven summits above 1500 feet, separated from one
another by trifling depressions. Again, the closeness of
the contour-lines on the north-east side is evidence of steep
slopes, but once the 1500 foot line is reached from the
north, there is usually a long interval before one meets it
again on the south side. A visit to the ground only
serves to deepen the impression derived from the map.
Lammer Law, though second in height, is the most
conspicuous summit in the Lammermuirs, and is that
most often ascended. Round it the denuding agencies
have been stronger or have had softer material to work
upon, for it stands well out both from the plateau and
from its neighbours.

Looking south and east from the summit one sees
rolling hills, covered with grass and heather, empty of

human interest, but filled with the charms of solitude. Away in the distance is a blue haze which means the valley of the Tweed, and beyond it the dim outline of the pastoral Cheviots. The sun is playing hide-and-seek with the hills as the fleecy clouds chase each other over the sky, and not a sound but the bleating of sheep breaks the summer silence. On turning to the north one is struck by a violent contrast. In the immediate fore-ground, indeed, are steep bare slopes furrowed by sharp-cut deans or dells formed by the hill-streams rushing Tynewards. But just beyond and below is the plain of Lothian, enclosed with trim hedges, fences, or stone dykes. Here and there a mass of foliage betokens the park round some country mansion, but for the most part the country is covered with rich grass, potatoes, or waving corn. The numerous tall chimney-stalks do not indicate factories, but the threshing-mills attached to many of the farm buildings. North-west one sees the pit-head works of Tranent and Prestonpans, elsewhere are many villages and hamlets, but no large towns. It needs no more to convince one that here agriculture is the chief occupation. Yonder are wide stretches of yellow sand marking the positions of Aberlady Bay and the Tyne Estuary. But generally the eye glances directly from the green fields to the blue waters of the North Sea or the Firth of Forth. Away in the distance are the hills of Fife, and beyond them the serried ranks of the southern Highlands.

Rising from the plain are several eminences whose abruptness speaks to their volcanic origin. They are

Traprain or Dunpelder (704 feet), North Berwick Law
(612 feet), and the Garleton Hills (590 feet). Traprain
is a dome-shaped mass on the right bank of the Tyne,
a mile and a half south-west of East Linton. The
Garleton Hills form part of a line of uplands on the left
bank of the Tyne extending from near East Linton right
to the west of the county and joining the Roman Camp

North Berwick Law

Ridge. On the summit is the Hopetoun Monument, a
tall, lighthouse-like shaft, which is seen from all over the
county. Their slope is especially steep on the Haddington side. But the most conspicuous of all is North
Berwick Law. It is easily the most striking feature in
the scenery of East Lothian. Of pure conical shape it
stands out 400 feet above the general level of its

surroundings, and has all the appearance of a miniature volcano. Like its neighbours, however, its existence in splendid isolation is due to the softer rock above and around having been removed by denudation. From its summit the view, including sea, hills, and plain, cannot easily be surpassed in Scotland.

A study of the relief of the county brings out several interesting points in regard to roads and railways. But the physical features of greatest human interest are the gaps between the Roman Camp Ridge and the sea, and between Dunbar and the Lammermuirs. It is very noticeable how road and railway come close together at both those points. Their strategical importance is confirmed by battles fought at each—Pinkie and Prestonpans at the former, and the two Dunbar battles at the latter.

5. Rivers and Lakes.

The hydrography of East Lothian is more interesting than important. The Tyne is the only stream which can be called a river, but its fame is based upon its trout and salmon, its picturesqueness, and its historical associations, not upon its size or its economic value. The watercourses may be considered under four heads : (a) the Tyne and its tributaries ; (b) those streams which enter the North Sea independently of the Tyne ; (c) the burns flowing into the Firth of Forth ; (d) the tributaries of the Whiteadder and the Leader.

(a) The Tyne, 23 miles in length, rises in Mid-lothian near Tynehead railway station at a height of

800 feet above sea-level, and for the first seven miles flows in a northerly direction. Then it crosses the boundary into East Lothian, about a mile above the village of Ormiston. A mile farther on is Winton Castle on the left bank, and then the Tyne separates the villages of East and West Pencaitland. The next important place is the county town, situated on a level haugh subject to

Old Bridge and Linn, East Linton

floods, one of the most disastrous of which occurred in October, 1775. Passing beneath the ruins of Hailes Castle, the Tyne reaches East Linton, beloved of artists. Here a sill of intrusive volcanic rock causes the river to form a cascade. The floor and sides of the miniature gorge are honeycombed with potholes, many containing at times the rounded pebbles which have excavated them. The fall is an obstacle to the passage of salmon, which are frequently to be seen in autumn in the pool below, occasionally making attempts to leap upwards. The total descent is about 20 feet. Passing the parish church of Prestonkirk, the river flows sluggishly in great bends through Tynninghame estate, and across the wide Tyne sands to the sea.

The numerous country seats with their well-wooded "policies" which line its banks combine to render the river highly picturesque. The magnificent timber which grows on these estates, the ornate pleasure-grounds, and the gently flowing current lend a southern aspect to the landscape which is somewhat foreign to the accepted idea of Scotland.

Of the tributaries, none of any importance come from the north. Practically the whole of the Tyne water is derived from the Lammermuirs. Their rocks are re-markably impervious, consequently the rain readily runs off the ground, and after a heavy fall the burns are very quickly flooded. Two right-bank tributaries only are worthy of notice—the Birns and the Coalstoun Waters. Both show a fan-like structure of affluents in their upper courses. Their deeply cut though narrow glens form no mean obstacles to the roads which run parallel with the hills. Active though these streams have been, those belonging to the Whiteadder and Leader systems are now more active still, for the watershed is close to the northern edge of the Lammermuirs. It is most probable that the rainfall is heavier on the south than on the north side, giving greater strength to the Tweed tributaries.

(b) About a dozen streams besides the Tyne enter

the North Sea from East Lothian. The most considerable are the Biel Water and the Dunglass Burn. The
musically-sounding Papana rises on Moss Law at a height
of 1300 feet. Papana is a curious name, but it may be
connected with Papple, a farm which the river passes a
mile below the village of Garvald. On entering Whittinghame it becomes the Whittinghame, and after travers-

Biel House

ing the grounds of Biel House, the Biel Water. The
Biel Water reaches the sea at Belhaven Bay. The Spott
Burn, a little south of Dunbar, is called the Broxburn
when it enters Broxmouth Park. Its mouth is a hazard
well-known to players on Dunbar Golf Links. *Broc* is
Gaelic for "badger," and the people of the little hamlet
outside the park walls will sometimes say they belong to

"Badgerburn." Dunglass Burn is remarkable for the depth of the gorge which it has cut before reaching the sea. The railway bridge across it is 125 feet above the level of the stream. All of those burns flow in well-wooded valleys, which contrast pleasantly with the rich though monotonous farm lands between them. These streams, in contradistinction to the Tyne tributaries, are gaining in extent at the expense of the eastern affluents of the Tweed.

(c) The West Peffer Burn, which flows into Aberlady Bay, has, like its companion the East Peffer Burn, been straightened and "canalised." Its most interesting feature is its name, which is Pictish. The other burns falling into the Forth, such as Seton Burn, are mere rills.

(d) Only the springs of the head waters of the Leader are in East Lothian ; but a considerable portion of the Whiteadder basin is in the county. The White-adder, according to local opinion, rises in the White Well at a little more than 1000 feet above sea-level ; but its "topographical" source is 200 feet higher and half-a-mile away on the slopes of Clints Dod. At first but a moor-land rill, it soon acquires volume, and has cut a deep valley through the Silurian strata. In several places the air-line distance between the thousand-foot contour is only half-a-mile, while to travel on foot from the one to the other would involve a descent and ascent respectively of 300 feet. The Whiteadder leaves the county a few hundred yards above the hamlet of Cranshaws. It is joined in East Lothian by the Faseny Water and the Bothwell or Bothal Burn, while the Monynut Water, rising on Bransly Hill, crosses the boundary two miles above its junction with the Whiteadder at Abbey St Bathans. All these streams have their banks studded with remains of castles and forts, an evidence that in the old days they provided routes for southern marauders.

The area of water in East Lothian is officially re-

Pressmennan Loch

turned at 418 acres, almost exactly two-thirds of a square mile. Of the dozen sheets of water—all with one exception small—most are reservoirs connected with the water-supply; some are ornamental lakes or ponds; one or two only are natural lochs, as Danskine. During and after the Glacial Epoch several large lakes must have existed similar to those in the Lothians farther west, but they have long disappeared. The little loch—about ten acres in extent—in Balgone grounds was formed by the late Sir G. Grant Suttie after vain and expensive attempts had been made to drain a morass. On an artificial island wild-duck have taken up their quarters. But the largest sheet of water in East Lothian is Pressmennan Loch, also artificial. It was created in 1822 by building an embankment across the eastern end of a valley which during the Ice Age must have been excavated by a powerful torrent, but which was then tenanted by a sluggish burn. The slope was so gentle that it was afterwards found necessary to dam up the western end also. The lake is nearly a mile and a half long and about 300 yards broad. Surrounded by a thick screen of magnificent trees it is one of the most famous of East Lothian's picturesque resorts

6. Geology.

Geology is the science that deals with the solid crust of the earth ; in other words, with the rocks. By rocks, however, the geologist means loose sand and soft clay as well as the hardest granite. Rocks are divided into two great classes—igneous and sedimentary. Igneous rocks have resulted from the cooling and solidifying of molten matter, whether rushing forth as lava from a volcano, or, like granite forced into and between rocks. Sometimes pre-existing rocks waste away under the influence of natural agents as frost and rain. When the waste is carried by running water and deposited in a lake or a sea in the form of sediment, one kind of sedimentary rock may be formed—often termed aqueous.   Other sedimentary rocks are accumulations of blown sand : others are of chemical origin, like stalactites : others, as coal and coral, are respectively of vegetable and animal origin.   For convenience, a third class of rocks has been made.   Heat, or pressure, or both combined, may so transform rocks that their original character is completely lost.   Such rocks, of which marble is an example, are called metamorphic.

A more scientific system of classification arranges rocks according to the date of their formation, and is based upon their fossils.   The organic remains preserved within the solid stone form a record which the geologist may read.   Igneous rocks, of course, contain no fossils, and in most metamorphic rocks the fossils have been destroyed or altered almost beyond recognition, but they can generally be "placed" by reference to the neighbouring formations.   If we find, for example, a basaltic dyke penetrating sandstone, it is usually a safe inference that the sandstone was there before the basalt.   On this system, then, the history of the earth's crust is divided into four epochs.   First is the Primary or Palaeozoic ; next the Secondary or Mesozoic ; third, the Tertiary or Cainozoic ; and fourth, the Quaternary or Recent. With the exception of alluvium, blown sand, boulderclay, and peat, which, as they form part of the surface covering, are technically "rocks," the whole of East Lothian belongs to the Primary or Palaeozoic Epoch, which comprises the following systems, the youngest being on top:

Palaeozoic Rocks

Systems
{
Permian
Carboniferous
Old Red Sandstone
Silurian
Ordovician
Cambrian
}

The East Lothian rocks, now to be described, belong to the Silurian, the Old Red Sandstone, and the Carboniferous Systems, along with their Intrusive Igneous rocks. The fact that nothing is left in the county representative of the Tertiary or Secondary Epochs is one of great importance. It is clear that during long ages the surface must have been exposed to the influence of the various denuding agencies, aërial and sub-aërial, and consequently the original features have been subjected to profound modification. Hence long-buried rocks have come to light.

The outstanding features in the geological history of East Lothian are (1) the deposition of the Silurian strata; (2) the mountain-making movements which contorted and elevated those Silurian strata; (3) the deposition of the Old Red Sandstone; (4) the formation of the Carboniferous system; (5) the subsidence resulting in the Great Rift or Central Valley of Scotland; (6) the various igneous intrusions; (7) the Glacial Epoch.

The Silurian rocks, with which may be included those belonging to the Ordovician system, cover practically the whole of the county south of the great boundary fault, that is, the East Lothian portion of the Lammermuir Hills. They consist of black and grey shales, greywackes, flags, conglomerates, and radiolarian cherts or impure flints. These rocks abound in fossils, the most characteristic being the graptolites. The continent whence these sediments—except the cherts which are of marine origin—were derived lay to the north-west, its coast-line extending from Connemara to Aberdeen. It was during the Silurian Epoch that vertebrate animals came into being These were fishes of humble form, but sufficiently developed to be preserved through long ages as fossils. Towards the close of the era there occurred outbreaks of volcanic activity, represented now by numerous but short dykes of igneous material running as usual from south-west to north-east, and by the bold triangular mass of Priestlaw Hill, an intrusive boss of granite.

These disturbances accompanied an upheaval of the earth's crust which must have covered a considerable area. The Southern Uplands of Scotland were then continuous with the Mourne Mountains of Ireland; and the contorted strata prove the existence of high ranges of mountains with deep valleys between.

Then followed a long period of denudation. At length a further subsidence took place favouring the formation of valley lakes. Into them flowed rapid rivers from the surrounding highlands, bearing large quantities of sand, while beaches of water-worn pebbles were formed on the margins. The climate too was dry, and thus the red sand was blown into the lakes to settle on the bottom, and be compressed into Old Red Sandstone. Curious

fishes swarmed in the waters, the most characteristic
being the *Holoptychius nobilissimus*. In our region a lake
extended from the Cheviots across the eastern Lammermuirs to a little beyond the present coast-line. The
remaining rocks of this series are on the south side of
the fault running from Dunbar to Gifford   Several
valleys which had been excavated in the Silurian strata
are completely filled with huge quantities of conglomerate.
The most remarkable mass is that crossing the Lammermuirs from Spott to Dirrington Law—a mass which must
have been not less than 2000 feet thick.

The close of the Old Red Sandstone epoch was
marked by subsidence of the land, and a return to humid
conditions. The fresh-water lakes became arms of the
sea and were greatly enlarged. Numerous rivers wandered
lazily over the flat country, here and there expanding into
marshes, and terminating in swampy deltas. The climate
resembled that of a hot-house. Life was abundant.
Luxuriant vegetation covered the whole land, quickly
growing and quickly decaying. Frequent subsidences
occurred, causing the rich carbonaceous matter to be
buried beneath layers of sand, mud, and gravel, or, where
the water was deeper and clearer, of limestone. The
pressure thus applied converted the wood into coal. Bed
after bed was laid down, each succeeding layer depressing
the masses below, and so preparing the way for further
depositions. Thus in the course of time a very considerable thickness of rock was formed, amounting in East
Lothian to more than 10,000 feet, although the water-
covering was never at any period very deep.

The Carboniferous Era was marked by great volcanic activity, accompanied by extensive faulting. This culminated in the event to which Scotland owes its present industrial prosperity. A large portion of the earth's crust, bounded on the north-west by a line drawn between Kintyre and Stonehaven, and on the south-east by a line drawn between Girvan and Dunbar, sank downwards. Thus was formed the Great Rift Valley of Central Scotland. From the margins on both sides streams flowed into the trough, bringing mud, sand and gravel, and protecting the precious carboniferous strata. The coal-fields of Scotland were thus preserved from denudation and lay hid till modern times. To the south-east of Dunbar another fault, roughly at right angles to the Lammermuir Fault, passes through Oldhamstocks, Innerwick, and Broxmouth. By it carboniferous rocks were let down to the level of the Central Valley, and thus a small portion of the Lowlands of Scotland extends a little round the corner, as it were, into the Southern Uplands, and cuts off that part of the Lammermuirs from the sea.

In various parts of the county, but especially on the Dunbar foreshore, are volcanic necks or vents filled with different materials. The Dove Rock at Dunbar is a plug of basalt; but many of the necks are filled with volcanic ash, and some with non-volcanic sedimentary matter. Dunbar Castle Rocks mark the site of another vent, choked with red and green tuff. St Baldred's Cradle, north of the mouth of the Tyne, is dolerite of a coarsely crystalline type. Of the volcanic rocks three are very important from the scenic point of view. The Bass Rock rises 350 feet above and extends to 60 feet below sea-level. It is a stock of trachyte, left in striking isolation and forming a conspicuous landmark at the entrance to the Firth. North Berwick Law is another mass of trachyte, towering more than 400 feet above the general level around. Traprain Law is a third trachytic intrusion,

Dunbar Castle Rocks

but of a different character. Its dome-like shape is interesting rather than picturesque. It was formed by a swelling of lava from below like a huge solid bubble, and it is known technically as a laccolite. Smaller and less perfect laccolites exist at Pencraig and Garvald. All the islands and skerries off the coast, such as Fidra and the Lamb, are of volcanic origin.

An immensely long period of quiescence followed, during which the forces of denudation lowered the level of the uplands, carved in them deep valleys, and pushed seaward the plains. The volcanic rocks being harder offered greater resistance than the softer sandstones and shales. Hence an approximation took place to the physio-

Ejected Blocks and Platform of Marine Denudation, near North Berwick

graphy of the present day—outstanding bosses and ridges of volcanic rock with valleys between, trending from west to east. Thick forest, now of a more temperate character, covered the whole region.

From causes that are still undetermined, the temperature steadily dropped, until on the Highlands and over the Southern Uplands great ice-sheets were formed. From

these mighty glaciers crawled slowly down over the Lowlands, overwhelming the tops of the highest hills, their
surfaces dotted with huge boulders, by their banks lofty
moraines, and pushing resistlessly before them great heaps
of debris. The Forth Glacier, as it may be named, must
have been at least 3000 feet thick; and, by its planing
work alone, it profoundly modified the face of the district.
The direction of its course was first from north-west to
south-east; but in East Lothian, diverted by the tributary
from the south, its course turned to the east and even
a little to the north of it. This is proved by the scratches
or glacial striae found on the rock surfaces at Thornton
Loch, Catcraig Quarry, the Garleton Hills, and elsewhere.

What, then, were the main results of the glacial
epoch? First, previously existing natural features were
accentuated: valleys were deepened, and consequently
masses of hard rocks became more prominent. North
Berwick Law and Traprain Law show their gentlest
slopes on the east, and are modified examples of "crag
and tail." Secondly, the ice-sheet covered the whole
region with a layer of clay mixed with stones, from
which most of the present surface soil is derived. This
till or boulder-clay filled up all the river channels. At
Beanston, near Haddington, a deep boring failed to pierce
the drift, which may mark a pre-glacial course of the
Tyne to the sea at Peffer Sands. The composition of
the till has sometimes given rise to curious mistakes. For
example, coal was once discovered at Oldhamstocks. An
examination proved, however, that the coal was merely

scattered through the boulder-clay, and had come from
the measures farther west. Thirdly, the glacier carried
on its surface boulders, which travelled for long distances.
When the ice melted the boulders were stranded far from
their place of origin. In some cases the term boulder is
almost a misnomer. Kidlaw, for example, is a mass of
limestone, a third of a mile long and a quarter broad.
Others are at Marl Law Quarry, and at Woodcote Park,
near Fala. On the foreshore at Dunbar are numerous
smaller masses of greywacke, which may have been trans-
ported from the Western Highlands. Fourthly, East
Lothian provides striking examples of the so-called dry
valleys—valleys which either contain no stream at all,
or are much too large to have been formed by the existing
rivulet. When the climate became milder, the retreating ice-sheet presented a precipitous front to the
Lammermuirs. The foot of this became a drainage
channel for the water from the hills and from the glacier
itself. The general trend of the channels is from south-
west to north-east. They are thus *parallel* to the Lammermuirs. One of them, three miles south of Dunbar,
has the very suggestive name of Dry Burn. Separating
Deuchrie Dod from the Lammermuirs is a channel 200
feet deep, cut in the solid rock. Another of great beauty
joins the Spott Burn just above Spott. When a post-
glacial stream entered a glacial dry valley the sudden
lessening of the slope gave rise to a delta which in most
cases modified the drainage. A typical specimen of this
" corrom " formation is the Aikengall Burn, which was
a pre-glacial tributary of the Oldhamstocks Burn, but

Aikengall Valley

(Glacial delta deposited by a tributary stream where it enters a dry valley)

now flows into the Braidwood or Thornton Burn. When
the main ice-sheet was retreating the climate was still cold
enough to permit of local glaciers on the Lammermuirs.
The violent torrents from these formed long banks of
sand and gravel, which, though not prominent from a
distance owing to the narrowness of the valleys between
them, are surprisingly revealed from close at hand. On
the final retreat of the ice-sheet lakes were formed and
deposits of silt made, but they were never so numerous
in East as in Mid and West Lothian. The clearest
remains are to be found near Broxmouth, at East Fortune,
and beside Balgone House. At the last a boring recorded
shell marl, glacial drift, and grey sand. In the shell marl
were numerous bones of red deer, wild boar and other
animals, as well as human skulls.

The surface soil is naturally derived from the under-
lying rock. In the more level parts of the streams are
haughs covered with alluvium, carried perhaps from a
distance. All round the coasts are extensive tracts of
blown sand, occupied by the golf-courses of Dunbar,
North Berwick, and Gullane. Soil from carboniferous
rocks conveyed by the ice-sheet overlaps the Old Red
Sandstone. But in general the surface covering is of
the same materials as the rock immediately beneath it.
Much of the soil in East Lothian is of excellent quality,
even tempting the plough to follow it well up the slopes
of the Lammermuirs. Behind Port Seton and Cockenzie
is a fertile strip possibly derived from the 100-foot beach,
which is occupied by market gardens. But the finest is
the soil derived from the Old Red Sandstone—Dunbar

red soil, stretching for fifteen miles lengthwise to the
Tyne.

The portions of the Carboniferous Series found in
East Lothian are as follows:

Lower Coal Measures.
Millstone Grit (Roslin Sandstone Series).
Carboniferous Limestone Series.
Upper Limestone Group.
Edge Coal Group.
Lower Limestone Group.
Calciferous Sandstone Series with contemporaneous
igneous rocks of various types.

The actual coal measures are confined to a small area
extending along the coast from Port Seton to Prestonpans
and inland past Tranent to Ormiston and Pencaitland.
The East Lothian coal-field is separated from that of
Midlothian by the Roman Camp Ridge.

7. Natural History.

Twice at least since the ice retreated, dry land has
joined Britain to Europe; and it was chiefly by means
of these bridges that immigrants came over to people the
desolate wastes. Naturally the hardier plants and animals
arrived first, and pushed close up to the line of the vanish-
ing ice. But, as the climate became more genial, more
delicate organisms followed; and then ensued a fierce
conflict for supremacy, which resulted in the present
more or less stable state of equilibrium. As the land

connection was not permanent, the number of species in this country must be less than in Europe, and the number will diminish towards the north-west, as being farthest from the source. Some species also had not time to cross, and others which did cross were unable to survive the altered conditions. Lastly, migratory birds are an exception to the rule, for many species visit Scotland in summer, and go no farther south.

The flora of the British Isles is classified under the four heads of (1) Alpine, (2) Sub-Alpine, (3) Lowland, (4) Maritime or Littoral. All four are represented in East Lothian. Since the Lammermuirs have a large area over the 1500-foot contour, several Alpine plants have lingered as relics of colder conditions. In the peat mosses *Rubus chamaemorus*, the cloudberry, grows freely, while *Juniperus communis* is occasionally met with. Heather is not so common as in the Highlands, many of the hills being green to the top; but sphagnum moss is found in the damper places. All over the uplands one sees the mountain pansy, but *Sisleria coerulea*, the blue moor grass, occurs rarely in the highest places. The deep dells descending from the Lammermuirs, particularly those facing the North Sea, are paradises for the botanist. Ferns are specially abundant. In Hall's Dean the *Cistopteris fragilis*, brittle bladder fern, and *Lastrea oreopteris*, mountain fern, are in profusion. Aikengall Dean provides a splendid show of *Polypodium dryopteris*, the oak-fern, with very large fronds, *Aspidium aculeatum* var. *lonchitioides*, and beautiful specimens of *Asplenium trichomanes*, the black spleenwort. Of other plants may be noted the guelder rose, with its globular snow-ball flower, the *Marchantia polymorpha*, bearing umbrella-shaped fruit, and the *Rubus saxatila*, stone bramble.

On and near the shore occur sand and salt-loving plants; as the glass-wort on Aberlady sands; *Silene conica*, the striated corn catch-fly, at Dirleton and Gullane; and *Chamagrostis minima*, the dwarf-grass, only at Gosford. An insectivorous plant found in wet places is the *Drosera*, the sun-dew. Near Aberlady church the *bifolius* variety of *Blysmus rufus* was discovered in 1894. The mare's tail and the deadly night-shade are not uncommon. In ponds occurs the bladder-wort.

The Bass Rock possesses at least one plant, *Lavatera arborea*, the red-flowering tree mallow, which is found nowhere else. Another almost equally rare plant on the Rock is the wild beetroot. The commonest of the Bass flowers are varieties of the *Silene*, or catch-fly, whose white or pink blooms splash the rock with colour.

Except for one branch there is little of interest to record in the fauna of East Lothian. Wild mammals are extremely rare, even the fox being discouraged, as there is no hunt nearer than Berwickshire. The county is renowned for its low ground and covert shootings, pheasants and partridges being reared in large numbers. In the season a pack of beagles finds sport among the hares. Adders are fairly numerous in the Lammermuirs. Badgers are occasionally met with, and are perhaps the most interesting survival of our ancient fauna.

Shore-birds abound on the level flats from Port Seton to Gullane, and there also the fowler makes large bags of migrants. The spring golfer at Luffness, in his occasional divergences from the "line" in search of a "sliced" ball, learns to his cost, by the beating of wings within a few inches of his ears, what it means to disturb a nesting tern.

In the lonely glens or over the bare hill-tops of the Lammermuirs may be heard the cries of the curlew and the golden plover Ring-ouzels haunt the burns, peregrine falcons prey on the grouse and rabbits, the pestilent carrion-crow is everywhere, while occasionally a pair of rough-legged bustards reign over the moors till they fall victim to the gamekeeper. Merlins also are not uncommon, and dunlins are known to nest. Herons are common all over the county and king-fishers haunt the reaches of the Tyne.

But it is the Bass Rock that we must visit to observe a bewildering wealth of sea-fowl. Only at Ailsa Craig or on the solitary stacks in the northern and western isles is bird-life as abundant. Nearly all the year round, but especially in the nesting season, every cranny has its occupant, the cliff is white with plumage and the air full of shrill or raucous cries. During the seasons of migration the lantern of the lighthouse proves an irresistible attraction, against which many rare visitants are dashed to death. Among the species recorded as nesting on the rock are the gannet, lesser black-backed gull, herring gull, kittiwake, guillemot, razor-bill, puffin, jackdaw, rock pipet, cormorant (common and shag), eider duck, peregrine falcon, and turtle dove. The last was probably accidental; and the peregrine falcon, to the regret of every ornithologist, no longer frequents the rock. The puffin, from

Sea Birds on the Bass Rock

Solan Goose on the Bass Rock

its comical appearance of mock solemnity, might be termed the clown of sea-birds. The blue or herring gull is called the robber, because it lives on what it can thieve from others. Yet the persistence with which it pursues the timorous guillemot, and, forces it to release its prey, which is then seized in mid-air, proves that the gains, although ill-gotten, are the result of hard work. The most interesting of all is the gannet or solan goose, which has for centuries been renowned as the chief bird of the Bass. From tip to tip of its outspread wings it measures at maturity—in its fifth year—full six feet. Some have been known to live for close on half-a-century. Until comparatively recent times its flesh was esteemed a great delicacy not only by natives of Scotland but also by English visitors. Before coming to the kitchen the bird was buried in earth for a few days, a process which to some extent mitigated its fishy flavour. Estimates as to its numbers vary from 100,000 to 5000, but an experienced ornithologist gave them in 1908 as from 7000 to 8000.

8. Along the Coast.

The coast-line, exclusive of islands and minute indentations, is thirty-two miles in length.

Coming from the direction of Edinburgh, we enter East Lothian when we cross, by a scarcely noticeable bridge, the Ravenshaugh Burn, a tiny rill which has, however, cut a deep dell in the escarpment, once a sea-cliff. For a few hundred yards the beach comes up to the strong retaining wall on the north side of the road. Outcrops of coal occur among the other rocks ; and, during the miners' strike in 1912, family parties might have been seen digging up the coal and conveying it home in perambulators. The little harbour close by took its present name of Morrison's Haven from Sir Alexander Morrison, who owned it at the time of the Union of Parliaments. The haven itself, formerly Acheson's Haven, dates much farther back. It was probably constructed by the monks of Newbattle, who not later than 1210 obtained a charter granting " insuper carbonarium et quarrarium in territoria de Travernent." The coal was shipped at this harbour. Quickly traversing the hamlet of Cuthill, we reach the straggling town of Prestonpans. The name of this old town comes from the older village of Preston, a few hundred yards inland. Here again we find evidence of the activity of the monks of Newbattle, who as early as 1198 were making salt at the pans of Althamer, now Prestonpans. Once clear of the town, the road runs close to the sea through Preston Links, no longer a golf course but often covered with fishing nets spread out to dry.

Presently we pass a pit of the Forth Colliery Company, part of whose workings extends beneath the firth. We then reach the combined village of Cockenzie and Port Seton. Here are two harbours, of which Port Seton is much the larger. In Cockenzie is the fine old mansion house of the Cadells. Outside its west wall ran the first railway made in Scotland. The rails were originally of wood. At Port Seton one would fain think oneself beyond the coal-mining district, but seams have been discovered, and soon this picturesque spot will be dis-figured with pit-head machinery, refuse bings, and smoke. At the east end of Port Seton we have before us a delight-ful stretch of coast-road curving like a sickle round Gosford Bay. On the right is a golf-course, one of the newest in the county.

Port Seton Harbour

On our left as we go eastwards is a stretch of links gay in summer with gorse and wild roses, with here and there a sheltered nook beloved of campers. Presently we skirt the high wall of Gosford Park. Behind it is a thicket of hawthorn and other trees, whose surface, level with the top of the wall, streams eastwards, a striking testimony to the force and frequency of the wild west winds.

Forsaking the coast for a little, we turn north-east and reach the village of Aberlady. We have passed on the left Craigielaw House, and beyond it the golf-course of Kilspindie. It is difficult to believe that Aberlady was the ancient port of Haddington. The bay is very shallow, and at low water the numerous fishing boats, which lie up here in the off-season, are high and dry. But we must

Fishing Fleet, Aberlady

remember that long ago vessels were usually beached at high water and unloaded when the retreating tide rendered them accessible to horses. The village now shows none of the signs of a bustling seaport, but it is frequented by the summer visitor who has no liking for organised "attractions."

The road skirts the head of the bay, passing the lovely

old mansion of Luffness, which dates back to the sixteenth century. Crossing the Peffer Burn, we traverse a wide expanse of links to Gullane. This is the beginning of a chain of golf-courses which collectively rival even St Andrews in fame. The first is Luffness. Farther on are the three courses of the Gullane Club. Gullane is one of the prettiest villages in East Lothian. Many of the buildings have white walls and red roofs, a green surrounded by houses lies a little off the main street, some fine trees adorn the west entrance, while above is the famous hill whose steepness every golfer exaggerates and whose breezes every golfer knows well. To the north is a broad beach of clean white sand.

A little past the station is Muirfield, the home of the Honourable Company of Edinburgh Golfers. Archerfield House, with its private links extending down to the shore opposite Fidra Island, is farther east. Then down the hill we come into Dirleton, the loveliest village in the county.

Two miles of road, uninteresting save for having the Law always in view, bring us to North Berwick. The Law towering above, the frowning Bass away yonder, the rocky coast fringed with islands, the town with its bright shops and prosperous look, attract its many visitors. The West Links is *the* famous golf-course. On the first tee Princes and Grand-Dukes, statesmen, millionaires, and quite ordinary people of all ages and both sexes await their turn to start. The very names of the holes are better known, to the caddies at least, than those of historic battlefields. The town itself rather leans to the aristocratic. Quality Street has a pleasantly "genteel" sound. The ruins of the old nunnery are carefully tended, while the fishing industry is comparatively small and is not obstrusive.

Three miles east is Canty Bay. Here the old red sandstone terminates in massive cliffs, on one of which stands Tantallon Castle. From the head of the Bay a

Golf Links, North Berwick

motor-boat conveys the visitor, but only in fine weather, to the Bass. The only access is on the side nearest the land. Near it is a ruined chapel, the remains of the famous fortress, and a modern lighthouse. Through the rock from east to west runs a natural tunnel 170 yards long and 30 feet high. It can be visited at low water. Hugh Miller compares it to the dark and dreary cavern into which Sindbad the Sailor was lowered along with his dead wife. The channel between the Bass and the mainland is as much as 75 feet deep. St Baldred, the patron saint of the rock, has had his name given to many features in the district, such as St Baldred's Cradle, and St Baldred's Boat.

At the mouth of the Peffer the rocky coast gives place to wide sands, which continue nearly to Dunbar. Turning inland past Whitekirk, we traverse Binning Wood to cross the Tyne by its lowest bridge at the hamlet of Tynninghame. A mile beyond we reach the London Road near the beautiful grounds of Nunraw and Biel House. Soon we skirt the wide sands of Belhaven Bay and enter the old town of Dunbar, which is second only to North Berwick as a summer resort. Its castle, dating at least from the days of Malcolm Canmore, though now a mass of almost unrecognisable ruins, was for centuries the greatest stronghold in Scotland and the key to the Lothians. Westwards are the fine Old Red Sandstone cliffs with here and there an isolated stack, and on the east are the links extending nearly to Catcraig.

The road leaves the town beside the modern Parish Church, whose square turreted tower is a conspicuous landmark, and skirts Broxmouth Park. Here were fought the two battles of Dunbar. For some miles the coast, though rocky is low, but gradually, as we approach the boundary, the Lammermuirs come nearer the sea. All the way from Dunbar road and railway keep close together, but a little beyond Linkhead they become contiguous, showing that the coastal plain has become very narrow.

Crossing the miniature gorge of Bilsdean, we arrive at the more deeply cut dell of the Dunglass Burn, where our journey ends.

9. Raised Beaches, Coastal Gains and Losses.

The evidence of various upheavals of the land, though not continuous, is very clear in several parts of the county Often, however, the traces are obscured by the presence of blown sand. East of Prestonpans sandy deposits about the 100-foot contour indicate a former beach at that level. At Prestonpans it is largely covered with market gardens. Near Gullane it reappears and is continuous and extremely obvious from the village to Collegehead, a mile eastwards, where it vanishes altogether. Authorities suppose that a huge mass of ice—a relic of the glacial epoch—covered the hollow from Aberlady to the Tyne, that the terrace was cut in the ice, and so disappeared when the ice melted. From here to the boundary no unmistakable evidence of a 100-foot beach remains. Where the coast is precipitous one would not expect it, and elsewhere one must postulate ice, blown sand, or subsequent removal by denudation.

The 75-foot beach is defined even less satisfactorily. It can clearly be seen on the face of the slope due north of Gullane. Elsewhere, as at Castleton and Seacliff, both near Canty Bay, sandy deposits indicate a beach, although no terrace is visible.

Fortunately, the 25-foot beach may be easily followed. From the Midlothian boundary eastwards its upper edge is marked by a bluff, eroded by the waves in the boulder-clay. From Aberlady to North Berwick it is less continuous, but there it reappears, and is most conspicuous at the mouth of the Peffer Burn. One may follow it for some way up the Tyne estuary. Hence it is not well marked till at Belhaven what was a low cliff, is now a gentle slope, 20 feet above the sea. Part of Dunbar south of the old harbour is built on it. From Dunbar to the Berwickshire boundary the beach extends almost uninterruptedly. Owing to the general low elevation of the coast, this raised beach is of less economic importance than in many other parts of Scotland, but the coast road from Levenhall to Gosford runs along the old terrace.

Erosion is proceeding between Port Seton and Kilspindie, a marked encroachment of the sea having been observed since 1892. This part of the coast faces west and is therefore opposed to the prevailing winds. At Dunbar, south of the old harbour, gardens and in some cases houses are just above high water mark. Every south-easterly gale causes damage. Groynes and a sea-wall have been built, but these require continued attention and repair. It is not at all unusual in winter for the houses to be bombarded by breakers. Near Skateraw in Innerwick Parish the quarrying of lime along the foreshore has led to some erosion.

No artificial works for the promotion of accretion exist in the county. Any growth of the land seawards, herefore, is due to natural causes. The Tyne and the

Fidra Island

numerous burns cannot be said to contribute much. They are too small, and in their lower courses for the most part too sluggish to convey a great amount of detritus to their mouths. We may take it, then, that Aberlady Bay, Belhaven Bay and other wide and semi-terrene expanses owe their sands to the transference of sea-eroded material by tides and currents.

Lighthouses are fairly numerous and well-placed. The mariner approaching from the south sees, as his first East Lothian guide, the light from Barness, built in 1899, three miles south-east of Dunbar—not always an effective warning, as occasional wrecks show. A beacon marks the dangerous shallow ending in St Baldred's Boat. On the south-east side of the Bass Rock a powerful light, erected in 1907, points out the deep channel between the Rock and the mainland. This channel is frequently used by vessels, but chiefly in calm weather and in daylight. The lighthouse on Fidra indicates the most north-westerly corner of the coast-line. All the havens are marked by pier-head and other guiding-lights.

10. Climate.

North-western Europe is peculiarly favoured in the matter of climate, as compared with other regions in corresponding latitudes. Proximity to the ocean is always an advantage, as is seen in the milder climate of Nova Scotia compared with the inland country due west; but when to that proximity are added warm currents of air and water from the mid-Atlantic, then the advantage is, in a sense, out of all proportion. Suppose the earth to rotate in the other direction, one result would be that the higher parts of Scotland would become permanently ice-covered like the interior of Greenland. We gain, however, not only warmth, but also abundance of moisture ; in some cases, considering the nature of the soil, a superabundance. As the highest land is close to the west coast, precipitation is greatest there, and it diminishes with fair regularity towards the east. A rainfall map constructed on the basis of using deepening shades or colours to denote heavier precipitation would correspond very closely with an orographical map, where deeper shading or colouring denotes increasing height of land.

On the average our country is not greatly favoured by sunshine. Even in summer with its long days, the amount of sunshine is smaller than in the Channel Islands, where the possible quantity is less. And in winter the deficiency is much greater than can be accounted for by merely astronomical reasons. The cause is of course the mountainous character of the surface, which encourages the formation of clouds.

A study of an isothermal map of Scotland reveals some interesting facts. In winter there is a general equality of temperature all down the west coast. In other words, Cape Wrath is, in January, as warm as the Mull of Galloway. Farther east is an approach to continental conditions ; and we find small areas, for example one round Aberdeen, where the temperature is less than 38° Fahr. The tendency is therefore for temperature to

decrease from west to east.   In summer the influence of
the land becomes stronger, and the isotherms are roughly
parallel to the lines of latitude.   Temperature decreases
from south to north.   At all seasons, however, the ocean
must be reckoned with.   In winter it warms, in summer
it cools the land exposed to it.   Isothermal maps, again,
are constructed on the ideal basis of imagining that the
land is all at sea-level.   An average allowance of one
degree for every 280 feet must be made.   For example,
the summit of Ben Nevis is about 16° colder than is
shown on the map.   Under winter conditions, when
snow covers the hills and not the lowlands, this difference
becomes even greater.

With the exception of rainfall, few statistics are
available for an account of the climate of East Lothian ;
but there is no reason to believe that the conditions
differ greatly from those of Midlothian.   Reasoning from
general principles, we should say that East Lothian is
colder in winter and rather cooler in summer.   The
contributory causes are that the county as a whole is
more elevated than Midlothian ; that it is nearer the
North Sea ; and that, though open to the west, it is less
open to the south-west, while on the south rises the wall
of the Lammermuirs, and the lowland portion, especially
the coast from the boundary to Canty Bay, is very much
exposed to the south-east.

The two stations from which we have temperature
records are Smeaton House near East Linton, 100 feet
above and four miles distant from the sea, and Haddington, centrally situated and 240 feet above sea-level.   The

average annual temperature at Smeaton is 47ʹ3°, at Haddington 46ʹ7°; that is, allowing for difference of elevation, the same isotherm would pass through both. The monthly averages are :

	J.	F.	M.	A.	M.	J.	J.	A.
Smeaton	37ʹ7	38ʹ2	41ʹ0	46ʹ3	50ʹ5	56ʹ7	59ʹ4	58ʹ1
	S.	O.	N.	D.	Year			
	53ʹ9	46ʹ8	40ʹ9	38ʹ2	47ʹ3			

	J.	F.	M.	A.	M.	J.	J.	A.
Haddington	37ʹ9	38ʹ8	39ʹ8	44ʹ6	49ʹ3	55ʹ3	58ʹ1	57ʹ0
	S.	O.	N.	D.	Year			
	53ʹ6	47ʹ0	41ʹ1	38ʹ4	46ʹ7			

July is the warmest and January the coldest month in both cases. As might be expected, Smeaton is warmer than Haddington in summer. In September, 1906, Haddington recorded a maximum of 88° and Smeaton 86°. A minimum of 5° occurred at Smeaton on January 28th, 1906. This gives an extreme range of 83°, while the average annual range is at Smeaton 21ʹ7°, at Haddington 20ʹ2°.

The prevailing wind in the county is from the west. Second to it are those from the easterly quadrant. The spring east winds are very severely felt, especially along the North Sea coast. Occasionally in winter wild south-easterly gales, which come lashing across the sea, make life miserable, render navigation dangerous and seriously hinder the local fishing industry. The sea-fog known as the easterly " haar " is very prevalent in spring and early summer. The average annual amount of sunshine at the Blackford Hill Observatory, Edinburgh, is 1388·1 hours, or 31 per cent. of the possible. One would expect

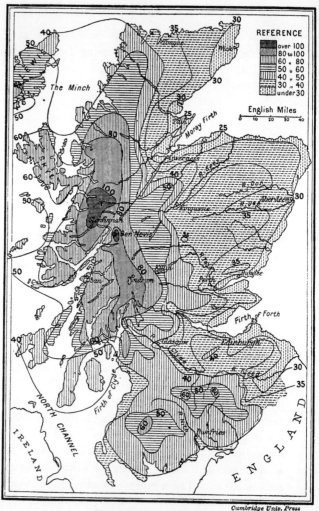

Rainfall Map of Scotland

(*By Andrew Watt, M.A.*)

East Lothian to be sunnier ; and indeed there are strong grounds for believing that Gullane is the sunniest spot in Scotland.

The rainfall of the county is naturally less than farther west. The map shows that the higher parts of the Lammermuirs have, on both sides of the watershed, a fall of more than 40 inches per annum. A narrow region along the east and north foot of the Lammermuirs, and including the village of Gifford, has between 30 and 40 inches. Nearly the whole of the remainder of the county has from 25 to 30 inches. A small patch in the north, extending from Dirleton to Gosford, and comprising Gullane and Aberlady, with less than 25 inches, is remarkably dry. Only one other part of Scotland, the head of the Moray Firth near but not including Inverness, has a similar record, and, judging from the figures for a number of years, the Gullane area has a slight advantage.

Every golfer who visits Gullane has observed showers passing down the Firth or some distance inland while around him it is dry and even sunny. This is particularly the case with thunderstorms, which contribute so notably to summer rainfall. For once, popular opinion, as the map shows, can be justified by statistics. But we do not agree with the popular explanation that North Berwick Law attracts the rain. The reasons probably are that the area is on the whole flat ; that local storms naturally tend to follow water such as the Firth of Forth, and valleys like that of the Tyne ; and that Gullane is almost as far as possible from the west, but so far sheltered from the North Sea as not to get the full benefit of the moisture coming from the east.

Taken as a whole, the climate of the low ground is ideal for agriculture, of the high ground for pasture, while round the coast the summer conditions of comparative drought combined with sea breezes are a powerful attraction to the visitor.

11.　People—Race, Dialect, Population.

As the Glacial Age came to a close, Neolithic man crept northwards to Scotland on the tracks of the reindeer, and established himself all over the lowlands. These primitive people were small, dark-haired, and long-skulled. They are generally named Iberians, Iverians, or Silurians, and their present-day representatives are the Basques, in the south-eastern corner of the Bay of Biscay.

After the Iberians came the great Celtic invasion. Being hunters and fishermen, the original inhabitants could never have been very numerous, and they were doubtless completely driven out, or at best overwhelmed. The scantiness of the evidence has made the identity of the Picts a subject of acute controversy. But the balance of opinion leans to the theory that the Picts were Welsh or Old Britons. As we shall see, there are in East Lothian several Celtic place-names which are not Gaelic but Welsh. It is extremely improbable that the Old British kingdom of Strathclyde ever extended so far east. Contemporaneous with it there existed the Pictish kingdom of Manau, which covered both shores of the Firth of Forth but did not reach far inland. Comparing the

PLACE NAMES
OF
EAST LOTHIAN

Based on the O.S. map

Scale of Miles

0   1   2   3   4   5

Gaelic   □
Pictish or Old British   •
Unknown   +
St Germains is French
All other names are Teutonic

place-name map in the present volume with that for West Lothian (*Linlithgowshire*, p. 56), one is struck with the distribution of the names whose origin is akin to Welsh. The evidence, then, is fairly strong that the Picts were Old British. The tribe which was paramount in our region was known as the Ottadini.

It does not appear that the Romans have left any traces of their presence. Not so the Angles, who spread northwards and westwards through the Lothians in the fifth and sixth centuries. There is a marked predominance of English names in the county, but this clearly decreases as one goes westwards. In 960 the Angles surrendered to the Scots the district between the Esk and the Avon ; while in 1018 by the great victory of Carham Malcolm II gained the whole province. Nevertheless the Angles though beaten were not driven out, and they gradually imposed their language on the inhabitants.

The study of place-names, while not strictly a part of geography, requires a knowledge of geographical conditions. A theorist often evolves a beautiful derivation, which a visit to the ground shows to be totally inapplicable. For original work in this field it is essential to have a command of the various languages, or to have access to a recognised authority[1]. To give a complete list of the place-names in the county would require a treatise to itself. But with the help of the general principles here stated, it will be easy to make classified

[1] The author again desires to offer grateful acknowledgments to Dr Watson, Professor of Celtic in the University of Edinburgh, for invaluable aid on this point.

lists. Two important exceptions must be made. The first is what may be styled "fancy" names. A man builds a house and calls it Bellevue or Drumnadrochit. These must be eliminated, and, in fact, they are indefensible. Such is Phantassie at East Linton, which in a deed of about 1800 is spelt both with F and Ph. Bleau's map (1654) calls it Trapren, and Phantassie does not appear in the "Retours" or the "Great Seal." We conclude, therefore, that it is from the French "fantaisie," and is an invented or fancy name. The other exception is that of names the derivation of which we have been unable to trace. These are all shown on the map and marked "unknown." Aberlady is a very puzzling name. The earliest spelling is Aberlessic. In 1328 it was Aberleuedy. *Aber* means mouth or confluence, but the name of the stream close by is Peffer, which is Pictish and of course ancient. The only name that may be Norse is Fidra, locally pronounced Fithera. A document of 1509 makes it Fetheray, the termination of which looks like the Norse word for an island. On the other hand it seems unlikely that the natives would use foreign words for a place-name. St Germains, the name of a mansion near Seton, is in Bleau's map, and is probably due to the French influence during the period of the Franco-Scottish Alliance (1285—1560).

Allowing for these exceptions, practically all of the names may be divided into three classes :

(1)   Old British or Pictish, i.e. Brythonic.

(2)   Gaelic.

(3)   English.

(1) Distinctively British elements are *tre*, "house," *aber*, "mouth" or "confluence," and *coed*, "wood." They are represented in the county by Tranent, "valley-stead"; Traprain, "house by the tree"; and perhaps Trabroun; by Pencaitland and Keith; while the first two syllables of Aberlady are certainly Old British. Other Brythonic names are Cairndinnis, Peffer, (Castle) Moffat, Pressmennan (*Manau* again), Prora ?, Carlaverock, Tyne, and possibly Papple and Papana.

(2) Gaelic was introduced after the conquest of the region by the Scots, and the disappearance of the Picts of Manau. Certain prefixes are characteristic of Gaelic such as *bal*, "stead," *drum*, "the ridge," and *kil*, "church." Examples are Ballencrieff, "tree-stead," Drem, "the ridge," Kilspindie, "the church or cell of St Pensandus." Other Gaelic names are Belhaven = *baile na h-aibne*, "village on the river"; Brox(mouth), "badger"; Garvald, "rough stream"; and perhaps Lammermuir, which may mean "big bare surface."

(3) English names are best classified according to their suffixes, and are very frequently personal. The most instructive suffixes are *ton*, "homestead," *law*, "hill," *ing*, "son of," and *wic*, "dwelling." Of the numerous examples, we may mention Haddington, Meikle Says Law, Clerkington, North Berwick, and Innerwick.

According to the Census of 1911 East Lothian has a population of 43,253. Thus, though twenty-fifth in area, it stands nineteenth in population, and eleventh in density. The increase over 1901 was 11·9 per cent., a rate excelled only between 1811 and 1821, when it was 13·1 per cent. In the near future East Lothian will probably become one of the few Scottish counties having more males than females. At present the deficiency is only 361. The three causes at work are the increasing development in the mining industry, the absence of textile manufactures, the great demand in Edinburgh for domestic servants, and the diminishing use of female labour in agriculture.

Of the 24 parishes, 11 show an increase and 13 a decrease. Of the latter all except Gladsmuir are purely agricultural parishes, and many of them are very small in area. The largest increases are in Ormiston, 34·3 per cent., Prestonpans, 39·6 per cent., and Tranent, 41·9 per cent., in all of which coal is king. Dirleton parish has advanced by 14 per cent., which is entirely due to the rapid rise of Gullane as a summer resort.

The population of 43,253 may be divided into inhabitants of municipal and police burghs, 20,302, and inhabitants of rural districts, 22,951. But there are several villages connected with mining, such as Ormiston, Pencaitland, Elphingstone, and Cuthill, which, along with the outlying suburbs of the burghs themselves, reduce the number of really rural inhabitants. Of the seven burghs five show an increase and two a decrease —Dunbar and East Linton. Of the five increasing, Tranent with a rise of 58 per cent. is the most remarkable. It should be noted that since 1901 the boundaries of East Linton, North Berwick, and Tranent have been extended.

As regards density of population East Lothian has 162 persons per square mile, slightly above the average for Scotland, but eleventh in order of the counties. Lanark has 1633, Midlothian 1373, while Sutherland has only 10.

12. Agriculture. Main Cultivations, Stock, Woodlands.

According to the returns for 1912, out of a total land area of 170,971 acres, 111,714, or about 65 per cent., are under crops and grass. The arable land amounts to 89,606 acres, leaving 22,108 acres for permanent grass. Woods and plantations occupy 10,777 acres; mountain and heath-land for grazing, 38,982. The remaining 9498 acres consist of waste land and that occupied by buildings.

The farmers of the Lothians have for long been celebrated for their skill and progressiveness. They have not rested content merely with the advantages derived from the natural fertility of the soil, and proximity to the great market of Edinburgh ; and they are recognised as the foremost in Scotland, which means the world, for a readiness to introduce new methods, and to utilise the discoveries of experimental science. It was not always so. John Cockburn, the last of his race to own Ormiston, although an exile in London, where he died in 1758, kept up his interest in his estate and wrote frequent letters to his gardener, Charles Bell. In one he says of Bell's father, who was a farmer, that "his husbandry goes no further than to gett bad grain one year and worse the next." Cockburn introduced proper drainage and regular rotation of crops, planted many trees, and was the father of scientific market and fruit gardening in Scotland. Another benefactor was Rennie of Phantassie, who was the first to bring the shorthorn into the Lowlands. Andrew Meikle of East Linton was the real inventor of the threshing machine. In 1712 Fletcher of Salton set up at West Salton the first barley mill in Scotland. At Hallhill near Dunbar Dr Hamilton in 1784 introduced the Swedish turnip, *ruta baga*, locally known as " baigies."

Mr A. G. Bradley in *The Gateway of Scotland* calls the Dunbar red lands " the cream of the county, probably the cream of the earth." The soil is composed of red loams of maximum fertility combined with friable easy-working qualities. It receives lavish and liberal treatment. Here the potato is king. Sometimes the yield is eight quarters per acre. The average yield per acre in Australia or the United States would be ruthlessly ploughed under in East Lothian. Not an uncommon rental is £4 or £5 per acre. In the county there are no fewer than 17 farms with an annual rental of over £1000. One tenant who holds three farms pays the princely rent of £4321 per annum.

The total area under cereals and pulses is 37,934 acres —7658 under wheat, 13,736 under barley, and 15,976 under oats. Small fruits such as strawberries and raspberries account for 336 acres ; orchards cover 74½ acres. Prestonpans has its market-gardens. Each of the large houses, and they are many, has its pleasance and its orchard. Behind Dunbar are well-known nurseries.

Potatoes occupy 9015 acres ; turnips, swedes, and mangolds, 14,604 acres. No less than 26,492 acres, or about 15½ per cent. of the whole, are given to clover, sainfoin, and grasses under rotation. As already mentioned, 22,108 acres are under permanent pasture, making in all, 28 per cent. of the available land devoted to the feeding of animals.

Although cultivation has crept a long way up the slopes of the Lammermuirs, on the north side at least, these hills carry a large number of sheep. The total for the county is 130,848, giving about 770 for each 1000 acres, as compared with about 1200 for Roxburghshire. They are nearly all of the black-faced variety, not nearly so valuable as Leicesters or even Cheviots, but hardier and more inured to stress of weather. Of cattle there are over 11,000. Horses for all purposes number 3771 ; pigs, 2289.

Haddington is the agricultural centre, and has one of the most important grain-markets in the Lowlands. Between it and Dunbar are the largest agricultural farms. For size of farms Berwickshire with 218·4 acres comes first, East Lothian with 203·1 acres, second. Over all Scotland the average is 85·9, in Shetland it is 16·2, in Sutherland 12·3. As regards yield per acre, taking a period of ten years, East Lothian is for wheat fourth, barley third, oats third, potatoes eighth, and turnips fifth, among Scottish counties. Wheat, barley, and oats average from 41 to 44 bushels per acre, potatoes about eight tons.

Woodlands occupy the relatively large area of 10,777 acres, or about one-sixteenth of the whole, which is a high proportion considering the large extent of the tree-less uplands. Among the most notable is the Binning Wood at Tynninghame, planted in 1707. As the site was a bleak and barren-looking moor, the neighbours were minded to scoff, but the experiment was very successful,

Binning Woods

and the wood is one of the finest in Scotland. Oaks are the predominant trees ; and especially in summer, a ramble along the numerous paths and drives is an experience of sheer delight. From the public road to one of the gates extends a magnificent avenue of lime trees. Close by are the equally charming woods of Newbyth. In Bleau's map of 1654 woods are shown at Keith, Humbie,

Ormiston, and Winton. These still flourish. In sheltered Ormiston are fig trees which produce ripe fruit in most seasons. These and the other trees in the wood were planted by John Cockburn. His beech trees still survive, and the yew which was already old in his time is to-day as vigorous as ever. Another famous yew stands in the park at Whittinghame, probably the oldest and finest in Scotland. The circumference of the spread of its branches as their ends lie upon the ground is considerably over 100 yards. In the grounds of Yester near Gifford are some magnificent beech trees of great girth and height. In 1892 the highest was 95 feet above the ground, and the most massive at five feet up had a diameter of 15 feet 6 inches. Close by are limes and Spanish chestnuts close on 100 feet high. But the most remarkable trees in the county are the six specimens of *Pinus pinea*, planted in 1846, on the railway embankment at Dunglass. These stone or umbrella pines are natives of the Mediterranean and no other specimens exist in Scotland. Finally, we may note the woods of Biel House, near Dunbar, and of Coalstoun House, near Haddington.

13. Industries and Manufactures.

East Lothian is not a manufacturing county. Most of the industries are of such small economic importance that in Lanarkshire they would not be mentioned. There is a distillery at Glenkinchie near Pencaitland. Belhaven and Haddington have each a brewery ; Prestonpans has two, with more than local fame for light table beer. At Prestonpans also are the works of the Scottish Salt Company. The monks of Newbattle, who started salt-making at least as early as 1198, of course evaporated sea-water, but that is now employed only as part of the process of manufacture from imported rock-salt. The fire-clay associated with the coal measures is worked up by a firm of potters, who also use china clay from Cornwall for making table ware. In the burgh are two soap-works. At West Barns near Dunbar are the premises of the British Malt Products Company, who manufacture horse-food. Lime is burnt at Oxwellmains on the battlefield of Dunbar, and at Harelaw near Longniddry. Saw mills are situated at East Linton, Pencaitland, Ormiston; flour-mills at East Linton and Haddington. North Berwick and Haddington have coach-builders, while the latter naturally has a maker of agricultural implements and a tanner. Cockenzie and Port Seton have boat-builders. Near the ruins of Gladsmuir Church is another ruin called Society, beside which is a very deep well. This was one of the breweries belonging to the Society of Edinburgh Brewers, founded in 1598.

Probably the most interesting of all is the woollen factory at West Mills, Haddington. It is the successor of a mill which was the parent of the modern industrial system in Scotland. Towards the close of the seventeenth century several acts were passed with the object of promoting Scottish industries. These measures attracted capital from across the border, and an Englishman, Sir James Stanfield, in 1681 issued to the public "A memorial

Haddington, from the West

concerning the cloath manufactory." A company was formed. The government did their best by prohibiting the export of wool and the import of cloth, but this protective measure could not be enforced. Labour also was scarce, imported skilled workmen were expensive and, despite severe repressive tactics, refractory. The final blow was the murder in 1687 of the founder by his own son, who suffered the extreme penalty; and the company was wound up in 1712. The trial of Philip Stanfield is interesting on account of the full employment of all the old-fashioned means of justice, including torture and the touching of the corpse by the accused.

14. Mines and Minerals.

East Lothian has the distinction of possessing the earliest-worked coal measures in the world. The first known English charter, that of Newcastle, is dated 1234; but between 1202 and 1210 Alexander de Seton was one of the signatories to a charter granting permission to the monks of Newbattle to work the Tranent coal. Very probably he himself was, previous to 1202, digging the coal at his own door. Some time ago a great subsidence took place on the railway close to Seton plantation. There was then discovered a seam of fine coal 8 feet thick and 6 feet below the surface. "The stoops were barely $4\frac{1}{2}$ feet wide, and the rooms or working-places only $3\frac{1}{2}$ feet wide, while, instead of wood-props, the upper part of the coal had been beautifully arched in order to support a brittle clay roof. The ancient miner had done all his excavations here on the 'in-gaun-ee' system, and one of the original openings was still quite traceable." Numerous other indications show that coal-mining is a long-established industry. A shaft has recently been sunk close to this seam at St Germains, and operations have also been begun nearer the village of Port Seton. Here, as at Preston Links farther west, the measures are prolonged under the Firth.

The East Lothian coalfield is separated from that of Midlothian by the Roman Camp Ridge and therefore lies in a basin. The coal has the advantage of being near the surface; the strata are frequently horizontal, and are little interrupted by faults or igneous dykes. In 1909 coal was raised to the value of £345,131, but great progress has been made since. The highest coal, the Great Seam, about 8 feet thick, has been largely worked out except under the sea. The Parrot seam usually contains a band of cannel or gas coal, which is hard and does not soil the fingers. The best household coals are obtained from the lower portion of the Splint coal and the Rough or Kailblades seam. The total thickness of the coal seams in East Lothian is about 35 feet.

The greatest activity at present is round Tranent. A serious drawback is the lack of a suitable harbour for export purposes, Cockenzie being quite inadequate. Nearly all the coal is sent by rail to Leith, and this causes delay and extra expense. A good miner can earn nine shillings a shift, which with nine shifts in a fortnight gives him over £2 a week. The miners live well, although

extravagantly, but on the whole are very steady-going.
Until recently the great hobby was whippet-racing, and
almost every miner had his dog. For some reason, however, the fashion has changed, and homing pigeons are
now *de rigueur*.

Contrast this with the conditions of little more than

Cockenzie Harbour

a hundred years ago. Then the miners and their families
were serfs. They were not at liberty to leave their birth-
places, and children were compelled to follow the occupation of their parents. The shafts of the mines were
circular, lined with stone, and fitted with crazy wooden
ladders. Up these toiled women and girls laden with
coal, pouring with sweat, sobbing and groaning with

distress. Yet, says an observer, the moment their sacks were empty they returned to the shaft laughing and even singing. It was not till 1843 that the employment of women underground was forbidden by act of parliament; and so recently as 1911 there died at Prestonpans a woman who remembered carrying coal up a neighbouring shaft. But it was not colliers alone who were serfs. The salters at Prestonpans were "thirled"; and the women were compelled to trudge with heavy loads to Edinburgh, and greet the burghers with the cry " Wha'll buy sa-at ? "

Seams of fireclay occur at Preston Links, Preston-grange, Northfield, Tranent, and Ormiston collieries, but are little worked. Blackband ironstone was formerly mined at Dolphingston and Penston. A haematite mine on the Garleton Hills near the Hopetoun Monument has not been worked for some years. Crystals of galena are visible in the limestone quarry at Catcraig near Dunbar, and in the dolerite dyke at Whitesands, but not in sufficient quantity to repay extraction. Copper veins in the granite of Priestlaw were formerly worked.

The county is not noted for building stone. Many houses in Dunbar are built of Old Red Sandstone, but it is not very durable. The best of the calciferous sandstones is the Bilsdean sandstone, of which Dunglass Mansion House was built a hundred years ago. The Edge-coal sandstone is quarried at Tranent but is not of great value. Of the igneous rocks porphyritic trachyte is quarried near Haddington, East Linton, and Dirleton, for building pur-poses. The trachytic phonolite of North Berwick Law, a rich red-brown in colour, is easily worked, and hardens

5—2

on exposure. North Berwick owes some of its picturesqueness to the use of this stone. Limestone is extensively worked at Skateraw near Dunbar, and Harelaw near Longniddry. About 30 to 40 tons a day are quarried at Skateraw, two-thirds of which is sent to the west of Scotland for smelting purposes. The remainder is burnt and makes excellent lime for agriculture. The limestone of Harelaw is used for building, plastering, agriculture, and gas-purifying.

Sixteen quarries in the county are now worked for road-metal. The best is composed of igneous rocks like the dolerite of Gosford Bay, the basalt of Kidlaw and Chesters, and the dolerite of Millstone Neuk. Silurian Greywacke and Carboniferous Limestone are also used but only locally. They are suitable for light traffic. Accumulations of glacial sands and gravels are worked at various places for local use, while building sand is obtained from a decomposed white sandstone near Winton Castle.

15. Fisheries. Shipping and Trade.

East Lothian has four fishing ports—Dunbar, North Berwick, Port Seton and Cockenzie, and Prestonpans. The third of these is the most important. In 1911 out of a total of £16,612 worth of fish landed, Port Seton and Cockenzie had £11,545: with 599 men and boys employed in the industry, leaving only 160 for the other three ports. To it belonged 130 vessels of an aggregate tonnage of 2793, the others had 28 vessels of 187 tons. At Dunbar and North Berwick the industry seems to be declining. Their principal fishing grounds are from one to five miles off-shore; and the chief kinds of fish are haddocks, codlings, and crabs, the last being the mainstay. The crab-catches indeed show an increase at Dunbar. At Dunbar also is the sole curing station in

Dunbar Harbour

the county. The Prestonpans boats have a similar range, but the catch is almost entirely codlings. Lines baited with clams, which are dredged by Port Seton and Cockenzie boats, are used for the codlings and haddocks, while crabs are trapped in creels. The beds at Prestonpans, which used to supply Edinburgh with the famous Pandore oysters, were ruined by over-fishing. At Port Seton and Cockenzie the industry is increasing. The boats range over the Firth of Forth and round the May Island. Besides clams, haddocks, codlings, and plaice are obtained. In winter herring visit the Firth, and are caught in nets. The herring is a very capricious fish. It sometimes comes in great shoals, at other times the shoals are small and few in number. In 1911 many shoals kept close in-shore so that from fear of damage to boats and nets they had to be left alone.

The home fishings, however, are comparatively unimportant in East Lothian. The Port Seton and Cockenzie fishermen depend mostly on following the herring to the principal Scottish and English centres. Some women and girls too accompany their relatives to help in mending the nets, and to work at the cleaning and curing of the fish. A family may thus return home in the autumn with a very respectable sum in hand to tide them over the winter.

Compared with the Solway, Tweed, Tay, and Aberdeen Districts, the salmon fishing in the Firth is of little importance. The total assessed rental for 1911 was £3756, while that of Aberdeen was just over £18,000. The salmon are caught in standing nets on the flat sands at the mouth of the Tyne.

The sea-borne trade of the county is very small. Dunbar exports corn, fish, and potatoes, and imports coal and timber. Prestonpans has a very limited export of coal, bricks, tiles, and salt. Were Port Seton and Cockenzie harbour to be enlarged and equipped with "handling" machinery, it would become the port not only of the pits close by, but also of Tranent and the whole coalfield.

16. History of the County.

Little is known for certain of the history of East
Lothian till the coming of the Angles. From 547 East
Lothian was English. But in 1018 by the great victory
of Carham, Malcolm II gained possession of the country
north of the Tweed, which has ever since been included
in Scotland. In the twelfth century Haddington had
a royal palace. Here William the Lyon sometimes
resided, and here was born his son Alexander II. In
1216 King John of England invaded Scotland and burnt
Dunbar and Haddington, then and, like all Scottish towns,
for long afterwards built of wood. In 1244 Haddington
was again burned. Despite these disasters the county
enjoyed great prosperity until the breaking out of the
War of Independence.

At the close of the thirteenth century the eighth Earl
of Dunbar, of the Cospatrick family, favoured the cause
of the English. While he was absent in England, his
wife handed over the castle to the Scots. Edward I sent
north an army under the Earl of Warrenne, which defeated
the Scots at Dunbar, April 1296. The day after the
battle, Edward arrived and received the surrender of
Dunbar Castle, which he restored to the Cospatricks.
There Edward II found refuge in his flight from Bannock-
burn to Berwick. At that period the county must have
suffered greatly from constant forays as well as by the
passage of invading armies.

During David II's minority, when Edward Balliol
backed by the English was a kind of shadowy king,

much fighting took place in the Lothians. In 1338
Montague, Earl of Salisbury, with a great army and
many engines, besieged Dunbar Castle. It was heroically
defended by "Black Agnes," Countess of Dunbar, and
after a siege of six weeks it was relieved by Sir Alexander
Ramsay of Dalwolsey. "Black Agnes" was a typical
Scottish heroine. When Montague brought forward a
"Sow," she called from the battlements,

> " Beware, Montagow,
>   For farrow shall thy sow " ;

and caused a huge rock to be let fall on the engine,
scattering its litter of pioneers. A few years later,
Eugène de Garancière arrived from France with men,
money, and arms to assist the Scots against the English
invaders. Edward III, however, ravaged Lothian, burning
Haddington, its monastery, and the famous church of
the Fratres Minores, which had attained such celebrity
as to be styled the "Lamp of Lothian." This raid,
in February 1356, was for long known as "the Burnt
Candlemas." In 1388 after the battle of Otterburn
Hotspur with 2000 men crossed the Lammermuirs, and
burnt Haddington with the hamlets of Hailes, Markle,
and Traprain. The spoilers, however, were routed and
deprived of their booty by an army from Edinburgh. In
1400 Henry IV re-asserted the old English claim to
suzerainty over Scotland, and with a large army passed
through East Lothian, the last time an English king took
the field in person against the Scots. He stayed in
Haddington; but for once no damage was done. Failing

to capture Edinburgh Castle, Henry re-crossed the border, after an invasion lasting fifteen days. A century later, 1503, Margaret Tudor spent a night in the Abbey at Haddington as she journeyed north to marry James IV.

After Flodden the English were too exhausted to invade Scotland ; but East Lothian, in common with all Scotland, suffered severely by the battle, many of the leading landowners as well as common people being slain. Thirty years later occurred the first of the two invasions of the Earl of Hertford. Haddington and other towns were burned to the ground. In 1547 Hertford, now Earl of Somerset, again invaded Scotland, and destroyed the castles of Dunglass, Innerwick, and Thornton. The army passed by Dunbar, Tantallon, and Hailes as too strong, and went on to their great victory of Pinkie.

To the horrors of foreign invasion were added those of religious persecution. In February 1546 George Wishart, who, attended by John Knox, bearing a two-handed sword, had been preaching at Haddington, came to Ormiston Hall. The house was surrounded by Both-well, who captured Wishart and conveyed him to David Beaton the Archbishop at Elphingstone. Shortly afterwards Wishart was burned at St Andrews.

Next year the English were firmly established in East Lothian. Lord Grey crossed the Lammermuirs above Yester, took its castle, and occupying Haddington, fortified it and left Sir James Wilford as governor with 2000 infantry and 500 horse. A mixed army of French and Scots, amongst whom were Highlanders, besieged the town, having the Abbey as headquarters. Next year

the French built a fort at Aberlady, the old port of
Haddington, in order to prevent the English fleet from
landing provisions or reinforcements for the garrison.
On July 17th, 1548, a Parliament was held at the Abbey,
when the French alliance was renewed, and the marriage
of Mary to the Dauphin agreed to. Except for the defeat
with great slaughter of a large English foraging party no
important incident occurred till the capture of the governor
when on a raid to Dunbar in 1550. Thereupon the town
was taken and the fortifications razed.

Mary Queen of Scots frequently visited East Lothian.
Seton Palace was one of her favourite resorts. Two days
after the murder of Darnley she arrived there with a large
retinue, and shot at the butts with Bothwell. After the
match the party adjourned to Tranent, and had dinner at
a tavern. In 1567 Mary was "captured" by Bothwell
and conveyed to her own castle at Dunbar, of which she
had put Bothwell in charge. After her marriage to him
she returned to Dunbar, from which she led her forces to
the miserable affair of Carberry Hill on June 15th, 1567.
From Carberry Bothwell fled to Dunbar, and thence
across the seas.

In 1650 was fought the second battle of Dunbar.
Cromwell, having been forced to retire from Edinburgh,
passed by Haddington to Dunbar, making his headquarters
at the house of Broxmouth. He was followed by Leslie,
who fixed his camp on the Doon Hill. Leslie had 27,000
men ; Cromwell, although he had shipped his sick from
Musselburgh, had "a poor, shattered, hungry, discouraged
army," of about 11,000 men. Leslie had cut off Crom-

well's retreat by occupying the dean of Dunglass to the south. Obviously he had nothing to do but to "sit tight," and await Cromwell's surrender. But Leslie was overborne by the clergy; the Scots descended from the hill,

Plan of the Battle of Dunbar, 1650

were attacked by the English, and in two hours put to utter rout. The English lost only thirty men, while of the Scots 3000 were killed, and 10,000 captured. Cromwell wrote to his wife, "The Lord hath showed us an exceeding mercy."

During the Covenanting times, the Bass Rock was used as a state prison. There were confined in dank and dreary dungeons about forty Covenanters, including Robert Gillespie, Alexander Peden, and John Blackadder. Blackadder had held a conventicle on the hill above Whitekirk, and he died on the Bass, aged 69, after a rigorous imprisonment of four years.

In 1690 sixteen young Jacobite officers, imprisoned on the Bass, surprised their guards, and bade defiance to the government of William. They were besieged for four years, and being compelled finally by superior force and hunger to capitulate, obtained favourable terms by treating their captor to biscuits and wine, thus making him think that they had plenty of provisions.

The Union negotiations of 1707 brought two East Lothian men into great prominence. Fletcher of Salton and Lord Belhaven and Stenton were by far the ablest supporters of Scottish Independence. Their speeches rang throughout Scotland, and, although they failed in their direct purpose, they helped to keep alive that feeling of distinct nationality which survives to this day.

In the afternoon of September 20th, 1745, two armies faced each other near Prestonpans. Sir John Cope had marched from Haddington, and Prince Charlie from Duddingston, near Edinburgh. The armies did not yet come to close quarters, for Cope was in a strong position, with back to the sea, and front and flanks well protected. During the night Robert Anderson of Whitburgh, a farm on Costerton Water, informed Lord George Murray and the Prince of a path through the marsh on

Cope's left. Through the darkness stole the Highlanders past the farm of Rigganhead, and in the mist of the morning burst upon Cope's army. The whole affair

Lord Belhaven

was over in less than half-an-hour. The royalist dragoons fled; but their leader, Colonel Gardiner, remained and tried to rally a body of infantry. He was shot and cut

down, under the walls of his own house of Bankton, and,
after the battle being carried off the field by his servant,
died that night in Tranent Manse.   The victory of the
Jacobites was absolutely complete.

17.   Antiquities.

Prehistoric relics in general are divided into the Stone,
the Bronze, and the Iron Age.   The Stone Age is composed of two periods : the Palaeolithic, marked by chipped,
though otherwise unworked, flint implements ; and the
Neolithic, by more highly finished and sometimes polished
implements.   No traces of the presence of Palaeolithic
man have been found in East Lothian.   Neither are
Neolithic relics at all numerous or varied.   Any signs
of dwellings, long cairns or graves have been obliterated
by cultivation.   Implements of flint and other stone have,
however, been obtained, particularly from the shore near
Gullane and Archerfield, and near the mouth of the
Tyne, less frequently from the hill country, as at Harelaw.   These consist of arrow-heads, borers, scrapers,
knives, and saws of flint and axes of other stone.   As no
flint occurs *in situ* in the county, we have evidence either
of primitive commerce, or, more probably, of periodical
visits to the chalk regions farther south.   A few stone
axes have been unearthed at Garvald, Stobshiel, and
elsewhere ; but these may quite easily belong to a later
period.

Relics of the succeeding Bronze Age consist of cairns,
stone circles, standing stones, hut circles and small cairns,

graves, kitchen middens, weapons, and implements. The cairns were erected for sepulchral purposes on hill-tops, spurs, or, if in the low country, often in prominent positions. Some of those which have been examined have contained stone cists with articles of bronze and

Piece of Pottery from midden at North Berwick

urns. They are circular in shape, and the largest of them, situated near Tynemouth, is 60 feet broad and 11 feet high. Other good examples are on Harestone Hill, and Whitekirk Hill. One stone circle survives in the "Nine Stones," only eight now, near Johnscleuch.

Of single-stone monuments, the most famous is Lot's, or Loth's, stone on Traprain Law. Here, according to tradition, was buried King Lot, grandfather of Kentigern. Another stone in the parish of Spott has on one face three cup-markings.

Stone Cist at Nunraw opened and emptied

Near Yester, Pencaitland, Garvald and elsewhere, short stone-coffins have been found containing either doubled-up skeletons or urns full of ashes. The relics in the cairns and coffins, and in the middens at North

Berwick and Gullane, include fragments of pottery, a bronze dagger with gold mountings, two spearheads, two swords, and a ring. The piece of pottery illustrated on page 79 bears distinct impressions of grains of wheat. An expert is of opinion that the wheat was of good quality. This find proves that wheat was cultivated in Scotland during the Early Bronze Age.

The coming of the Iron Age with the gradual increase of population demanded means of defence more elaborate than before. Its characteristic feature was the forts, remains of which are found all over the country. They almost invariably occur in groups, the individuals being separated sometimes merely by a few hundred yards of ordinary ground. The forts on the south side of the Lammermuirs were undoubtedly designed for the defence of the routes over the hills northwards, but at least some of the forts on the north side, as those near Gifford, are not on any line of communication.

The shape of hill-top or knowe-top forts is generally governed by the contours, the others are circular or oval. The defences consisted of walls of stone and earth. On the top of the walls was a palisade of wood probably strengthened with turf. At Harelaw, a small portion of one wall gives evidence of vitrifaction. Some forts have as many as four walls, others have none at all where cliffs or other natural defences existed. Foundations of hut-circles are numerous in some of the enclosures. A curious feature of these forts is the absence of a water-supply within their bounds, although a stream is not infrequently close at hand. More than thirty forts are scattered over

the county.   The most interesting are at Harelaw, Stobshiel, Kidlaw, Friar's Nose near Priestlaw, and the
Chesters near Drem.   Traprain Law was once a fortified
hill, and it is one of the largest of such structures in the
east of Scotland.   According to tradition it was the
residence of King Lot.   Without doubt, it was of great
strength and was the home of a relatively large number

Traprain Law
(*Lines of ancient fortifications accentuated*)

of people.   The extreme measurements of the works are
1100 by 300 yards.   The ancient name of the hill was
Dunpender or Dunpelder, which may mean "stockaded
fort."   The more modern name is probably derived from
that of a farm upon its flanks, but both names are old
British.

Remains of later times consist chiefly of cemeteries
and kitchen middens.   The most notable early cemeteries

so far discovered are at Belhaven, Lennoxlove, Innerwick, and Nunraw The contents of the middens are shells, bones, and pieces of pottery, glazed and modelled on the wheel.

At Coalstoun House is preserved in a silver casket a pear which was a wedding gift from Sir Hugo de Gifford, the wizard of Yester, to his daughter. The story goes that the fortunes of the family hang upon the pear being kept uninjured, and that this has been tested twice. The tradition is well authenticated, and the pear is probably six centuries old. It is shrivelled but quite recognisable.

18. Architecture—(a) Ecclesiastical.

We may here neglect the buildings of the early Celtic Church, and begin with the church architecture that came north from England. Saxon and Norman influence was first felt in the eleventh century, and during the next two hundred years there was a gradual development of the Romanesque or Norman style. This was marked by the rounded arch and tower, and the introduction of ornament. From the thirteenth to the sixteenth century the Gothic or Pointed architecture was supreme This style had three stages—the First Pointed in the thirteenth century, the Middle Pointed or Decorated in the four-teenth and fifteenth centuries, and the Third or Late Pointed period. While the Perpendicular or Late Pointed prevailed in England and the Flamboyant in France, in Scotland there arose a style peculiar to itself. Its chief characteristic is the barrel vaulting covered with over-lapping stone slabs, which obviates the use of wood to support the roof. During the sixteenth and seventeenth centuries the Renaissance influence slowly penetrated, but

Tynninghame Church

it competed with the Tudor style from England and the surviving Gothic.

Tynninghame Church is now used as a mausoleum of the Haddington family. The original edifice dedicated to St Baldred, who died in 606, has completely disappeared.

What remains is pure Norman with elaborate chevron ornaments and billet-and-hook mould. The west end of the choir, the chancel arch, and the pillars of the eastern apse are well preserved. A rather uncommon feature is the two arched recesses for altars. The church of St Andrew at Gullane is much more ruinous, but the now built up chancel arch is Norman. It was dedicated to St Andrew in the twelfth century, bestowed on Dryburgh Abbey early in the thirteenth century by Sir William de Vaux, and in 1446 erected into a collegiate institution by Sir Walter de Haliburton. Additions were made at the Reformation, but in 1633 it was abandoned and a new church built at Dirleton. About a mile east of Haddington are the ruins of St Martin's Church. The chancel arch of the chapel is late Norman. The old walls are of irregularly coursed brown freestone, and the gables are lofty, perhaps to allow of an upper storey. A peculiarity is a number of holes about ten inches square, which may have held the beams of the scaffolding.

The pre-Reformation name of Prestonkirk was Lynton. Here was the second of the churches dedicated to St Baldred, the third, now completely gone, being at Auldhame. The only ancient part of Prestonkirk is the small eastern choir, now cut off from the rest of the present church by a solid wall. It is in the First Pointed style. Pencaitland church is still in use, but most of it dates back only to the sixteenth and seventeenth centuries. The chapel on the north side was once vaulted and covered with stone slabs.

Only one church, but that is of great beauty, re-presents the Decorated period in East Lothian, the parish church of Haddington. It was designed as a whole and completed in the middle of the fifteenth century[1]. The nave is the only part now roofed, and it is used as the parish church. At different times it has been repaired and improved. The original building was cruciform in shape with a total internal length of nearly 200 feet.

Haddington Church

The tower at the crossing is thirty feet square and ninety feet high and had originally a crown like that of St Giles, Edinburgh. A very beautiful feature is the main west doorway, a circular arch adorned with mouldings in Late

[1] Accordingly the *Lucerna Laudoniae* of Fordun and Major must have been the church of the Franciscans, a little to the north, of which no vestige remains.

Decorated style.   The choir and transepts are ruinous,
but the gargoyles and terminals are enriched with carvings
of grotesque animals and foliage.   On the north side of
the choir is the mausoleum of the Lauderdale family of
Renaissance work.

Several churches belong to the Late Pointed period
The Collegiate church of Dunglass, founded in 1403,
though used as a stable in the eighteenth century, is in
excellent preservation.   The roof is a pointed barrel vault
with heavy overlapping slabs of stone.   In the south wall
is a fine sedilia with three seats.   The church is not now
a place of worship, but in the south transept is the burial
ground of the Halls of Dunglass.   Seton Collegiate
Church is also disused except as a burial ground of the
Wemyss family.   The original plan was that of a complete
cross but the nave was never built.   It existed in the
fourteenth century, was rebuilt in the fifteenth and completed by the erection of the tower by the Dowager
Lady Seton in memory of her husband, killed at Flodden.
In 1544 the English burnt the timber work and carried
off the organ and bells.   The tower is crowned with
a broach spire, the top of which is unfinished.   Very few
examples of such spires exist in Scotland.   The north
and south end windows of the transepts are divided into
two compartments by heavy stone mullions built in
courses, and each compartment is filled with smaller
tracery.   St Bothan's Collegiate Church at Yester was
founded by Sir William Hay in 1421.   Its nave and choir
are not in the same line, and the walls are four feet thick,
thus requiring no buttresses.

Keith Church, now ruinous, was erected in the reign of David I as a private chapel by Hervie de Keith, King's Marischal. At the east end are two narrow lancet windows and a large vesica-formed opening above. The Winton aisle of Pencaitland church is pure Gothic of the fourteenth century, but the main body was built soon

St Mary's Church, Whitekirk

after 1660. The tower and portions of the west gable of Prestonpans church are of ancient date. From the louvred openings of the tower "Jupiter" Carlyle and his father were spectators of the battle of Prestonpans.

The gem of the county is, or rather was, the church of St Mary at Whitekirk. It is now, as the result of the dastardly outrage in February, 1914, a roofless ruin. The walls still stand, but all the woodwork of the interior, the historic Bible, communion plate, and interesting furniture of the Haddington gallery have been utterly destroyed. Portions of the church, notably the charming entrance archway, date from the fourteenth century. The red freestone embosomed in trees, surrounded by picturesque cottages, made a very pleasing picture.

19. Architecture—(b) Military.

The Normans were great castle-builders as well as great church-builders, and their influence is distinctly visible in the early fortifications of Scotland. After the War of Independence came a complete change. Strong square towers with flat and battlemented tops and arrow-slits for windows, were built in commanding situations. Gradually a court-yard was added enclosed by a wall. The wall gave place to buildings, stables, and living rooms. When more space was required, the early plan was to add another storey or two; later on, a wing was thrown out. Hence the L, T, E and Z plans of more secure days. Another development was the substitution of a balustrade for a wall, which is the best mark of the abandonment of military for domestic architecture.

Many castles were built in East Lothian, all of them probably on the model of the rectangular keep of the Normans. Of Dunbar Castle so little is left that it is impossible to describe its appearance in the days of its splendour. Dismantled by the Lords of the Congregation after Bothwell's flight in 1567, the castle was left

The Goblin Hall, Yester

a prey to the elements, till it was finally ruined in 1842 by the blasting operations when a sea-channel was made right through it for the Victoria harbour.

The oldest part of Dirleton Castle dates from the early thirteenth century. These are the south-west towers, of which two are round and one square, the lower part of the south-east tower, and the adjoining walls. Besieged by Bishop Beck in 1297, it was restored a hundred years later, and again in the fifteenth century; while in the sixteenth, the time when common halls were being given up, a new hall was erected for the private use of the owner and his friends. Although a ruin since 1650, when General Lambert battered it, the castle is most interesting not only in itself but also from its picturesque surroundings. Its situation in a well-kept garden beside the village green reminds one of Kenilworth.

Situated on a promontory above the junction of Hopes and Gifford Waters, Yester Castle commands an old route over the Lammermuirs. The walls remaining are 40 feet high and 6½ feet thick, defended on the third side of its triangle by a ditch 50 feet wide and 20 feet deep. This is crossed by a beautifully built bridge of undoubted antiquity. The Castle was erected by Sir Hugo de Gifford the Wizard in 1268, is now owned by the Marquis of Tweeddale, and has thus been in possession of two families for more than 700 years. Its most striking feature is the Goblin or Bo' Hall, an underground chamber of massive construction, reached from the castle by a sloping passage partly cut out of solid rock, and from about the level of the stream, by a concealed entrance. Since a well has been found,

besides signs of a fireplace, it must have been meant as
the last retreat should the castle be taken, and possibly as
a secret rendezvous for a sally. It is unique in Scotland,
but similar structures exist at Windsor, at Dover, and in
France. Fordun, writing about a century after the castle
was built, ascribes it to demoniac agency, and the tradition
persists locally to the present day. Hailes Castle belonged
in Queen Mary's time to James, Earl of Bothwell, and
was probably built by a Hepburn in the thirteenth century.
The north postern adjoining the donjon is very ancient.
A postern stair of great strength leads down to the river
Tyne. Although commanded by neighbouring heights,
it ranked, in pre-artillery days, with Tantallon, Dunbar,
and Dirleton, as one of the strongholds of East Lothian.

No castles were built in East Lothian between 1300
and 1400, but six belong to the following century. Of
these one of the best preserved is Elphingstone Tower.
It is remarkable for the number of rooms in the thickness
of the wall. One is a gallery 30 feet long and 6 feet
wide. Another provides a peep-hole into the great hall
by way of the fire-place. Whittinghame Tower is in
excellent preservation. The entrance doorway has the
Douglas arms over the lintel, recalling the fact that there
Morton, Patrick Douglas, Bothwell and Lethington met
to plot the murder of Darnley. The battlements are
high and quite entire. The ground floor of Preston
Tower is in two storeys. The lower was a dungeon and
could be reached only by a hatch from the upper, which
was a guardroom. A straight stair and also a hatch led
from it to the real first floor of the castle. The entrance

to the latter was by a movable wooden stair, which was completely commanded by a wooden platform above it. In the seventeenth century a two-storey house was built on the top of the tower. The architecture is quite different from that of the tower and gives a very quaint effect. Stoneypath Tower near Garvald is very dilapidated. It is on the L-plan, and the staircase is "as it were folded

Tantallon Castle

over and placed inside the re-entering angle" instead of being a square structure projected in it. In the hall is a double fire-place which was meant to serve both the hall and the kitchen behind it.

Last but most famous of the purely military castles is Tantallon. "Build a brig to the Bass, ding doun Tantallon" is a local criterion of impossibility. It was built about

1400. Beginning as a royal castle under the constableship of the Lauders of the Bass, it passed to Murdoch, Duke of Albany, then to "Bell the Cat," back to James V, and now it belongs to the Dalrymples. Surrounded on three sides by the sea, it had enormously strong defences on the fourth side. The road to it led across two ditches with mounds and outworks, then across a third ditch to the entrance, over which was a peel tower. The curtain walls were thick and high. A steep passage leads to the quadrangle, once completely enclosed by buildings, but those on the east side have disappeared, possibly undermined by the sea. In the centre of the courtyard is a well, 100 feet deep. In 1639 the castle was taken and damaged by the Covenanters, and in 1650 reduced to its present state by Monk.

20. Architecture — (c) Domestic and Municipal.

The sixteenth century saw the gradual abandonment of fortified castles in commanding situations for pleasant sites sheltering ornate and spacious mansions. Towers became turrets, battlements parapets, and curtain walls balustrades. Hence the Scottish baronial type of architecture—truly national if not truly artistic. Keith House was of this type, though it was repaired in Gothic style in the eighteenth century.

Winton House is the most beautiful seat in East Lothian, and one of the finest examples of Renaissance

King Charles's room, Winton House

architecture in Scotland. It was designed by William Wallace, the king's master mason. The notable features are the tall Elizabethan chimneys adorned with spirals, and slates on the roof, which are cut into patterns. Within are spacious rooms, in particular the drawing-room and King Charles's room, the ceilings of which are of fine plaster work of the time of James VI. Across the Tyne is Fountainhall, another fine old Scottish house in well-kept grounds.

A famous East Lothian seat is Lethington near Haddington. A Maitland bought the estate from the Giffords in the fourteenth century; but the present building dates from the second half of the fifteenth century. Built on the L-plan, it is spacious and ornamented with battlements, gargoyles and carved monsters. Adjoining the old tower and connected with it is the modern mansion. In the seventeenth century the name was changed. The then owner playfully offered to sell it to Lord Blantyre, well knowing he could not afford to buy it. Blantyre's daughter, however, was the Duchess of Lennox, a reigning beauty at the court of Charles II and the model for the figure of Britannia on our coinage. She provided her father with the money; and the next time the offer was made, to the owner's chagrin, it was accepted. Hence the new name of Lennoxlove.

Northfield House, Prestonpans, has a quaint hall on the first floor; Magdalen's House, locally named the "Barracks," is a fine piece of Renaissance work; Lord Fountainhall's House has a towering chimney and ingle neuk, while some of the fire-places have projecting hoods.

At Cuthill, near Prestonpans, several cottages have outside stairs. In Dunbar, near the harbour, is a house with two outside stairs, one above and at right angles to the other. Haddington House, dated 1680, has a notable entrance, porch and staircase, and a glass door of seventeenth century work. Dunbar Town Hall, dating probably from the sixteenth century, with its spire and crow-step gables, is extremely picturesque.

21. Communications—Roads and Rail= ways.

East Lothian affords some beautiful examples of the control of communications by natural features, and also to some extent of man's contemptuous treatment of physical obstacles. Observe how the roads converge upon the railway line at Prestonpans after being miles apart to the east; how road and rail cling to each other from Dunbar southwards to the boundary; and how from Gifford and Garvald two roads start to cross the Lammermuirs, only to meet on the Whiteadder. Elsewhere roads run a short distance into the hills, and then stop or return by some other way. On the other hand, from Haddington two routes cross boldly over nearly the highest part of the Garleton Hills, one to North Berwick, the other almost as straight as possible to the old port of Aberlady. Some of the curious "dry" valleys parallel to the Lammermuirs are traversed by roads, the higher of them crossing the little streams by fords.

Entering the county from Edinburgh, a traveller has the choice of three routes, the low, the middle, and the high. For Berwick-on-Tweed, he will climb the hill to Tranent, whence gentle undulations take him to the cobble stones of Haddington. Thence to East Linton the road runs above the Tyne along the flanks of the Garleton Hills. Crossing the river, it is level to West Barns and uphill into Dunbar. Or one may turn aside at Hedderwick and avoiding Dunbar go direct to Broxmouth. The coastal sill is followed to the boundary, where is a fine bridge over the Dunglass Burn. The middle road follows the railway, crossing and re-crossing it by bridges or on the level to Drem. It is a useful alternative to the coast-road, which, though flat and picturesque, is narrow and winding as far as Dirleton.

A glance at the O.S. map is sufficient to show that Haddington is a real centre for roads, although it is only a terminus on a branch line of railway. From each end of the street roads diverge north and south. Curiously enough, the bridges across the river do not carry main roads.

Very little is known about the ancient roads. In Bleau's map of 1654, the only highway given in the county comes from Musselburgh to Preston (not Preston-pans), and then goes inland, following roughly the line of the present Haddington branch railway, but avoiding the county town, to which it sends an off-shoot. Farther on it crosses the Tyne at "Linntyn briggas," and follows the existing road to Dunbar. Several drove roads cross the Lammermuirs. One, called the Herring Road, leads from

Dunbar over its Common on the hills to the Whiteadder. Another connects Innerwick with the Monynut Water and Abbey St Bathans in Berwickshire. A third leaves the old Gifford road and follows a very straight line to Longformacus.

On these roads were rude bridges—some of which still stand—just wide enough for a pack-horse. One of

Nungate Bridge

them called Edincain's bridge was at the Thornton Burn. Only a few traces of it remain. Both at Dunglass and Pease bridges the traveller on looking over will see remains which prove that at one time the road crossed at a much lower level. The most famous bridge is the Nungate at Haddington, possibly dating from the twelfth century. Less than twelve feet wide, it is extremely steep at both ends.

East Lothian is monopolised by the North British Railway Company. The main line, which is part of the East Coast route to London, enters the county above Levenhall, sweeps round by Drem to avoid the Garleton Hills, and proceeds by East Linton to Dunbar and the coast. Branches serve Aberlady and Gullane, Dirleton and North Berwick, on the one hand, and Haddington on the other. A light railway leaves the main line at Monktonhall near Inveresk, divides at Ormiston, sending one branch to Macmerry, and another to the charming country of Humbie and Gifford. It has long been proposed to extend the latter to Garvald. Several mineral lines serve the various collieries.

The only tramway line in the county is that which follows the coast road from Levenhall to Port Seton, a distance of $3\frac{1}{2}$ miles.

22. Administration and Divisions.

For judicial and administrative purposes, a sheriff-principal presides over the three Lothians. East Lothian has a sheriff-substitute resident in the county town. There are also honorary sheriffs-substitute. In Haddington meet the Justice of Peace Court, the Licensing Court, and the Licensing Appeal Court, as well as the County Committee for Secondary Education.

East Lothian contains 24 parishes: Aberlady, Athelstaneford, Bolton, Dirleton, Dunbar, Garvald, Gladsmuir, Haddington, Humbie, Innerwick, Morham, North

Berwick, Oldhamstocks, Ormiston, Pencaitland, Prestonkirk, Prestonpans, Salton, Spott, Stenton, Tranent, Whitekirk and Tynninghame, Whittinghame, Yester. There
are three royal burghs: Haddington, Dunbar, North
Berwick; and four police burghs: Cockenzie and Port
Seton, East Linton, Prestonpans, Tranent. The county
is divided into 27 electoral districts, which return 27
members to the County Council. There are two district
committees, one for the Western District with headquarters at Haddington, the other for the Eastern District,
centred at Dunbar. The Lord Lieutenant is the Earl of
Haddington and there are 23 deputy lieutenants. The
powers of the County Council are many and varied,
ranging from water supply and police to lunacy and
plague-prevention. Parochial matters are managed by
elected Parish Councils. Their chief duty is the administration of the poor laws. Part of the rate they are
empowered to levy is spent by the School Boards, elected
bodies responsible for education within their areas. The
County Constabulary is partly under the County Council
and partly under the Commissioners of Supply.

The military force is in the Scottish Command, with
headquarters at Edinburgh. The barracks in Dunbar are
a depot of the Royal Scots Greys. The Territorial
Association, established by Act of Parliament in 1907,
has for its President the Lord Lieutenant. The unit is
the eighth battalion, Royal Scots, composed of four
companies, stationed respectively at Haddington, Tranent,
Prestonpans and East Linton.

Until the passing of the Redistribution Act in 1885,

the burghs of Haddington, Dunbar and North Berwick joined with Lauder in Berwickshire and Jedburgh in Roxburghshire in returning a member of parliament. This arrangement has now ceased and the county as a whole elects a member.

23. Roll of Honour.

Several noble families have been intimately connected with the county, as the Innes-Kers, the Hopes, the Hays, the Setons, the Maitlands. The first Maitland to become famous was Sir Richard (1496–1586), a Judge in the Court of Session, and a writer of religious verse. It was after him that the Maitland Club was named. His collection of manuscripts was bought by Samuel Pepys and they are still in the library of Magdalene College, Cambridge. His eldest son, William, best known as Secretary Lethington, was one of the most prominent figures in Scottish history in the reign of Queen Mary. A younger son became Lord Chancellor of Scotland and Baron Thirlestane. The Chancellor's grandson was the notorious Earl of Lauderdale, who ruled Scotland in Charles II's time. Two outstanding statesmen at the time of the Union in 1707 were Andrew Fletcher of Salton and Lord Belhaven. A Judge of the Court of Session, Lord Fountainhall, belonged to the Lauder family. He wrote two volumes of legal decisions, and charming gossipy books entitled *Observes* and *Journals and Observations on Public Affairs from* 1665 *to* 1676. Sir Thomas

Dick Lauder (1784–1848) wrote several books, of which the best known to-day is *Scottish Rivers*. An eccentric

Sir David Baird

native of East Lothian, also a judge, was Lord Grange of Preston House. In 1732 he had his wife abducted by some Highlanders and conveyed to St Kilda, giving out that she was dead and holding a mock funeral. It was ten years before she was discovered, and then only by accident.

The county can boast of its soldiers. Sir David Baird, of the Bairds of Newbyth, served with distinction in India, in Egypt, at the Cape, and in the Peninsula. On Moore's death at Corunna Baird succeeded to the command. For his services he received the thanks of parliament on four occasions. Other famous soldiers were Colonel Gardiner, killed at Prestonpans; John, Earl of Hopetoun, who distinguished himself in the Peninsular War and whose monument crowns the Garleton Hills; and Thomas Alexander, Director-General of the Medical Department in the Crimea. A native of Athelstaneford, a Hepburn, served in the Thirty Years' War and ultimately was made a Marshal of France. In another sphere Peter Laurie, born in Morham of humble parents, became famous. He made a fortune in London and was Lord Mayor in 1832.

Among ecclesiastics belonging to East Lothian, John Knox stands supreme. Haddington and Morham both claim to have been his birthplace. One of the last and greatest of the schoolmen, John Major, was born at Gleghornie near Tantallon. He taught in the Universities of Glasgow, Paris and St Andrews, and died at St Andrews in 1550. Of his voluminous writings, the most interesting to us is his *History of Greater Britain*. Patrick Hamilton, John Knox and George Buchanan were all pupils of Major. Another ecclesiastic connected with East Lothian

by residence was Gilbert Burnet, Bishop of Salisbury and
historian, who became episcopal minister of Salton in

John Knox

1665. Although he soon went to London he never forgot
his first charge and in his will left £2000 for the education
and clothing of thirty poor children and the upkeep of the

library. The library still contains some books presented by the bishop. Of a different type was John Brown, who came to Haddington as an Associate Burgher minister. He is best remembered for his *Bible Dictionary*. One of his grandsons, Samuel, was a great chemist. Dr Alexander Carlyle of Inveresk was born in Prestonpans Manse. He is always known as "Jupiter" Carlyle, because he frequently sat to Gavin Hamilton, the sculptor, for the head of the god. His *Memorials* are rich in interest. Dr Caesar of Tranent was a fine example of the parish minister of the old school who survived into the present century. Robert Moffat, the great African missionary and father-in-law of David Livingstone, was born at Ormiston in 1795. In the manse at Gifford was born John Witherspoon, the first head of Princeton University in New Jersey, who was a member of Congress in 1776, and signed the Declaration of Independence. Robertson, the historian, was minister of Gladsmuir and wrote his *History of Scotland* in the manse. Later he was Principal of Edinburgh University and Historiographer Royal of Scotland. He died in 1793. Robert Blair, author of *The Grave*, was minister of Athelstaneford. His third son was Lord President of the Court of Session and a great judge. Blair's successor in the parish was John Home, author of *Douglas*, who died in 1808. A contemporary and neighbour of Blair and Home was Adam Skirving, writer of *Hey Johnnie Cope*, whose topical ballads made him locally famous. Another Skirving of Athelstaneford gained fame as a portrait painter. Samuel Smiles, author of *Self Help*, was born in Haddington. George Miller set up the first printing-press in East Lothian, in 1795. His *Cheap Magazine* was the pioneer of modern circulating literature. James Miller, his son, wrote a history of Dunbar, and the *Lamp of Lothian*. Carlyle had associations with Haddington through his wife, Jane Welsh, whose grave is in the choir of Haddington Church.

Three of the Scottish poets of the sixteenth century

Jane Welsh Carlyle

are claimed for East Lothian, with more or less probability. There seems no doubt in the case of William Dunbar, the greatest of the early Scottish writers, author of *The Thistle and the Rose*, celebrating the marriage of James IV and Margaret Tudor, of *The Dance of the Seven Deadly Sins*, and of *The Lament for the Makers*. Gavin Douglas, third son of Archibald, Earl of Douglas, " Bell the Cat," may have been born in Tantallon and was Rector of Prestonkirk. Poet, politician and ecclesiastic, he was Bishop of Dunkeld, and, as translator of the *Aeneid*,

> " in a barbarous age,
> He gave rude Scotland Virgil's page."

Sir David Lyndsay, Lyon King-at-Arms, author of *The Satire of the Three Estates*, was probably born in Garmylton (now Garleton) Tower. His poems did much to prepare for the Reformation and remained for two centuries the most popular reading in Scotland.

24. THE CHIEF TOWNS AND VILLAGES OF EAST LOTHIAN.

(The figures in brackets after each name give the population in 1911, and those at the end of each section are references to pages in the text.)

Aberlady (pa. 963) is a small village on Aberlady Bay, near the mouth of the Peffer Burn. (pp. 33, 39, 43, 44, 51, 54, 55, 74, 97, 100.)

Athelstaneford (pa. 666) is a village consisting mostly of one broad street, on the east slope of the Garleton Hills. The village green has a monument to Blair, author of *The Grave.* (pp. 100, 104, 106.)

Belhaven is a small watering-place in Dunbar parish. (pp. 42, 44, 55, 61, 83.)

Bolton (pa. 256) is a hamlet round the parish church. In the churchyard the mother, brother, and sister of Robert Burns lie buried. (p. 100.)

Cockenzie and **Port Seton** (2400) form a combined police burgh in the parish of Tranent. The chief industries are fishing and coal-mining. Cockenzie is Gaelic, "Kenneth's nook." (pp. 30, 31, 33, 37, 38, 44, 62, 65, 66, 68, 69, 70.)

Dirleton (pa. 2064), in Dirleton parish, is one of the most beautiful villages in Scotland. (pp. 33, 40, 51, 56, 67, 91, 92, 98, 100.)

Drem, a village in Athelstaneford parish, is the railway junction for North Berwick. (pp. 55, 82, 98, 100.)

Dunbar (3346), a royal burgh of David II, is a popular summer resort with bracing air, a fine golf-course, and picturesque rocks. The handsome parish church contains the elaborate marble tomb of George Home, Earl of Dunbar, who was James VI's Lord High Treasurer. (pp. 2, 3, 10, 13, 16, 23, 24, 28, 30, 42, 44, 46, 56, 58, 59, 61, 62, 67, 68, 69, 70, 71, 72, 73, 74, 75, 89, 92, 97, 98, 99, 100, 101, 102, 107.)

East Linton (877), a small police burgh in Prestonkirk parish is a busy and prosperous place. (pp. 3, 12, 14, 48, 54, 56, 58, 62, 67, 98, 100, 101.)

Garvald (pa. 511) is a small village, hidden in a sheltered hollow. Close by is Nunraw, a beautifully restored modern representative of the ancient nunnery. The present drawing-room, once the refectory, has a ceiling of oak planks nailed to the joists. It is covered with paintings of quaint design, including animals and musical instruments. Part of the ceiling is in the National Museum of Antiquities, Edinburgh, the remainder is still in place. Although the date is the early sixteenth century, the colours are fresh and bright. (pp. 25, 55, 78, 80, 93, 97, 100.)

Gifford is a pretty village in Yester parish. Yester Castle is renowned for its Goblin Hall, a cavern reputed to have been formed by magic art. Scott makes use of the tradition in *Marmion*, Canto III. (pp. 23, 51, 61, 91, 97, 99, 100, 106.)

Gladsmuir (pa. 1433) is a small hamlet in Gladsmuir parish. Another is Penston, inhabited by colliers. Gladsmuir means the moor of the " gled " or hawk. (pp. 56, 62, 100, 106.)

Gullane, in Dirleton parish, is a popular summer-resort with famous golf-courses. Gullane was once a noted training-ground for racehorses. (pp. 30, 33, 40, 43, 51, 81, 85, 100.)

Haddington (4140), the county town, is an ancient royal burgh and a fine example of mediaeval town-planning, where everything was subservient to church and market. One of the leading grain markets in Scotland, it contains the largest Corn Exchange out of Edinburgh. The Knox Institute (1878), the successor of the old Grammar School, is a flourishing centre of higher education. In the High Street is the Town Hall with a spire 170 feet high. (pp. 1, 27, 39, 48, 49, 55, 59, 61, 62, 67,

Public School, Tranent

71, 72, 73, 74, 76, 85, 86, 96, 97, 98, 99, 100, 101, 102, 104, 106, 107.)

Innerwick (pa. 676) is a pretty little village in Innerwick parish, in the extreme east of the county. (pp. 24, 44, 55, 73, 83, 99, 100.)

Longniddry, a village in Gladsmuir parish, is the railway junction for Haddington and Gullane. (pp. 62, 68.)

Macmerry, in Gladsmuir parish, is a village whose inhabitants work in the neighbouring collieries. The name may be Gaelic for "the merry or wanton one." (p. 100.)

North Berwick (3247), an ancient royal burgh, is now a popular summer resort, delightfully situated, with fine sands and splendid golf-courses, excellent bathing and a healthy climate. Behind the town rises the green cone of the Law, with an extensive

Whittinghame House

prospect from its summit, while out of the sea in front towers the Bass Rock. (pp. 3, 30, 40, 44, 55, 62, 68, 69, 81, 97, 100, 101, 102.)

Oldhamstocks (pa. 404) is a small village in the east of the county near the Berwickshire border. (pp. 24, 27, 101.)

Ormiston (pa. 1598) is a village in Ormiston parish, a mining and agricultural region. John Cockburn began his pioneer work in agriculture here; and Moffat the missionary was a native of the village, which has a monument to his memory. In the church is a brass with an inscription by George Buchanan commemorating his pupil Alexander Cockburn. (pp. 31, 56, 57, 61, 62, 67, 73, 100, 101, 106.)

Pencaitland (pa. 1273), a village in Pencaitland parish, has a fine old church and an ancient cross. The parish is rich in limestone and freestone. (pp. 14, 55, 56, 61, 62, 80, 88, 101.)

Port Seton. See **Cockenzie.**

Prestonpans (1923) is a burgh in the most densely peopled part of the county, with such industries as coal-mining, fire-clay-working, pottery, brewing, salt- and soap-making. The market gardens excel in producing cabbage and leek plants, and parsley. (pp. 3, 8, 11, 13, 31, 37, 43, 56, 58, 61, 62, 67, 68, 69, 70, 76, 96, 97, 98, 101, 104, 106.)

Salton or **Saltoun.** In the parish (386) there are two villages—East and West Salton. The parish church is at East Salton. (pp. 58, 76, 101, 105.)

Tranent (4369) is an important burgh in a mining and agricultural region. Coal was mined here early in the thirteenth century at least. The large school has over a thousand pupils and is a Junior Student centre. The same parish contains the hamlets of Elphingstone and Meadowmill. (pp. 11, 31, 55, 56, 64, 65, 67, 70, 74, 78, 98, 101, 106.)

Tynninghame is a small village beautifully situated on a gentle slope near the Tyne. (pp. 15, 42, 60, 84, 101.)

Whitekirk is a hamlet with an ancient church, destroyed by fire in 1914. The church was for several centuries a favourite place of pilgrimage. Its most distinguished pilgrim was Aeneas Silvius, afterwards Pope Pius II. (pp. 42, 79, 88, 101.)

Whittinghame (pa. 523) is a hamlet in the parish of the same name. Whittinghame House is the home of Mr A. J. Balfour. (pp. 16, 61, 92, 101.)

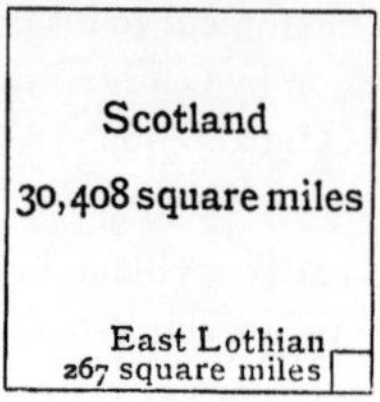

Scotland

30,408 square miles

East Lothian
267 square miles

Fig. 1. Comparative areas of East Lothian and all Scotland

Scotland

4,759,445

East Lothian
43,253

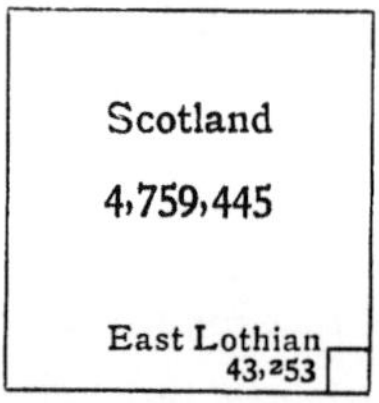

Fig. 2. Comparison in Population of East Lothian and
all Scotland

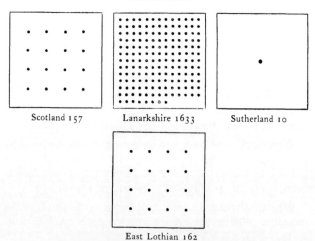

Scotland 157   Lanarkshire 1633   Sutherland 10

East Lothian 162

Fig. 3. Comparative density of Population to the
square mile in 1911

(Each dot represents ten persons)

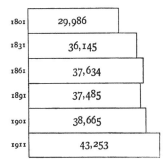

Fig. 4. Growth of Population in East Lothian

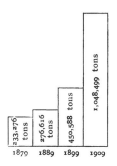

Fig. 5. Progressive output of Coal in East Lothian

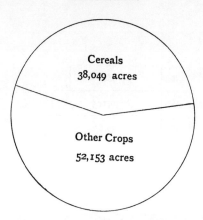

Fig. 6. Proportionate area under Corn Crops compared with that of other land in East Lothian in 1912

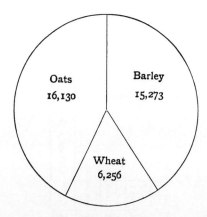

Fig. 7. Comparative areas under Cereals in East Lothian in 1912

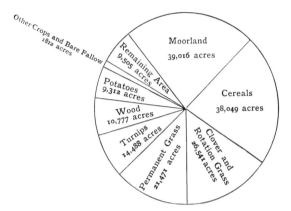

Other Crops and Bare Fallow 1812 acres

Remaining Area 9,505 acres

Potatoes 9,312 acres

Wood 10,777 acres

Turnips 14,488 acres

Permanent Grass 21,471 acres

Moorland 39,016 acres

Cereals 38,049 acres

Clover and Rotation Grass 26,541 acres

Fig. 8. Comparative areas of land in East Lothian in 1912

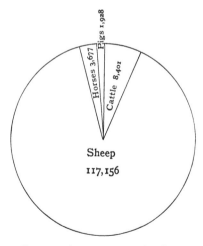

Pigs 1,928

Horses 3,677

Cattle 8,401

Sheep 117,156

Fig. 9. Comparative numbers of different kinds of Live Stock in East Lothian in 1912

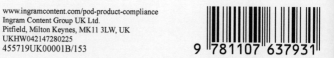